THE POLITICS OF TRASH

THE POLITICS
OF TRASH

HOW GOVERNMENTS USED
CORRUPTION TO CLEAN
CITIES, 1890–1929

PATRICIA STRACH
KATHLEEN S. SULLIVAN

CORNELL UNIVERSITY PRESS

Ithaca and London

First published 2022 by Cornell University Press

Library of Congress Cataloging-in-Publication Data

Names: Strach, Patricia, author. | Sullivan, Kathleen S., author.
Title: The politics of trash : how governments used corruption to clean cities, 1890–1929 / Patricia Strach, Kathleen S. Sullivan.
Description: Ithaca [New York] : Cornell University Press, 2022. | Includes bibliographical references and index.
Identifiers: LCCN 2022006272 (print) | LCCN 2022006273 (ebook) | ISBN 9781501766985 (hardcover) | ISBN 9781501766992 (pdf) | ISBN 9781501767005 (epub)
Subjects: LCSH: Refuse and refuse disposal—Political aspects—United States. | Refuse and refuse disposal—United States—History—19th century. | Refuse and refuse disposal—United States—History—20th century. | Municipal government—Corrupt practices—United States—History—19th century. | Municipal government—Corrupt practices—United States—History—20th century.
Classification: LCC HD4483 .S77 2022 (print) | LCC HD4483 (ebook) | DDC 363.72/850973—dc23/eng/20220716
LC record available at https://lccn.loc.gov/2022006272
LC ebook record available at https://lccn.loc.gov/2022006273

To the dedicated people who promote health in our communities—past, present, and future

Contents

ACKNOWLEDGMENTS

Garbage collection is such an expected pub-
lic service that it is hardly recognized in political science literature. We
chose to study it for precisely that reason. We wanted to examine a government function that is so basic that we could nail down the resources
that a government relies on to get its work done. What we found were
thousands of pages of descriptions of putrefying waste. We encountered stories of precarious living conditions, opportunistic political
actors, beleaguered public officials, and overlooked people who make
things work and keep the community safe. It has been a privilege to
spend time with the ingenious, committed, and diligent people who
cleaned up a mess so well that we don't even notice what a feat they
accomplished.

We found generosity from colleagues who have read our work and
fostered this book through many iterations, in particular Julie Novkov,
Carol Nackenoff, Eileen McDonagh, Ruth O'Brien, Tim Weaver, and
Bruce Miroff. We have received helpful comments and advice from
professional colleagues: Joe Bowersox, Dara Cohen, Richard Dilworth,
Richard Ellis, Jennifer Griffin, Kimberley Johnson, Ron King, Paul
Manna, John Meyer, Rob Mickey, Susan Moffitt, Colin Moore, Joel
Tarr, Michael Tesler, Jessica Trounstine, Alexis Walker, and the audi-
ences at UMass Amherst, the Policy and History Reading Group, Ohio
University's Political Science Department and Center for Law, Justice
and Culture, and the Rockefeller College Brown Bag.

We were guided by the archivists and librarians through their col-
lections and their respective cities at the University of Pittsburgh
Archives Service Center; Library and Archives Division, Historical Soci-
ety of Western Pennsylvania, The Senator John Heinz History Center;
City Archives & Special Collections, New Orleans Public Library; The
Historic New Orleans Collection, The Williams Research Center; The
Charleston Archive, Charleston Public Library; South Carolina Histori-
cal Society; San Francisco History Center, San Francisco Public Library;

Missouri State Archives; Missouri Historical Society; New York State Library; University at Albany Library; Ohio University Alden Libraries; and the Louisville Free Public Library.

We appreciate the support from professional associations, especially the Western Political Science Association and the Urban Politics section of the American Political Science Association, which have recognized our coauthored work. And we are grateful, too, for financial support from the University at Albany, State University of New York (FRAP A) and the Robert Wood Johnson Foundation Health Policy Scholars program.

We have received the most capable assistance from Katie Zuber, Elizabeth Pérez-Chiqués, David Trowbridge, Nora Sullivan, and Essa Dampha.

We are thankful for the support we received in turning this research into a book from Michael McGandy, the team at Cornell University Press, and the anonymous reviewers of our book. Related research has been published in *Statebuilding from the Margins: Between Reconstruction and the New Deal, Social Science History,* and *Studies in American Political Development.*

We appreciate the dedication of health professionals in the United States, especially the caring doctors, nurses, and staff at the James Cancer Hospital and Solove Research Institute.

We are fortunate to have had the support of our families. We are grateful to our spouses, Jerry Marschke and the late Steven Fetsch. We are grateful to David Wakefield. To our children especially, Harry Fetsch and Nora Sullivan, Joe Marschke and Lily Strach, who have spent formative years of their childhood privy to garbage discussions, we thank you for your patience, and have every confidence that you will clean up this world around you.

THE POLITICS OF TRASH

Introduction
The Garbage Problem

On garbage day, a resident of San Francisco places her food scraps, soiled paper, and plant debris in a compost bin because she has to; both composting and recycling are mandatory in San Francisco. But she would probably do it anyway; it is gratifying to minimize contributions to the landfill. The city fosters a "customer-oriented" experience, offering bin sharing, relationships with collectors, and education rather than harsh fines.[1] Reducing waste at the curb follows her into the home. She may reduce trash accumulation by not bringing excess packaging into her home in the first place. At the store, she forgoes single-use packaging by bringing her stash of containers and tote bags. In her kitchen there might be a countertop compost bucket and bins designed to sort garbage from recycling. Paper, nonwaxed cardboard, glass bottles, and other materials go in the recycling bin. Whatever is left (and it is not likely to be much) goes in the garbage can. A garbage truck emblazoned with *Recology*—not *City of San Francisco*— whisks the waste away, taking it to a facility owned by a company with a vision of zero waste. Recology traces its roots to the Sunset Scavenger Company, the company that has been picking up garbage in San Francisco since the late nineteenth century.[2]

Remarkably, garbage collection doesn't look a whole lot different than it did at the turn of the twentieth century. Even in San Francisco, where progressive practices are based on diverting waste from the landfills rather than just sanitary disposal, the removal of garbage has not really changed from when engineers first developed newfangled carts and bins and space-saving devices. Traveling around a city emptying curbside cans looks much as it did when a nationwide wave of municipal garbage ordinances was passed in the 1890s, and it's still done locally. A person who moves to a new town or city has to figure out when and how to put out their waste because cities continue to decide what they collect, when they do, with what trucks and cans, and where they dispose of it all.

Advancements made in the 1890s remain familiar today, but routine, effective trash collection and disposal service was not always the norm. Garbage collection tended to follow expected patterns. Before municipal garbage collection, householders were responsible for disposing of their own waste. They might bury waste in their yard or privy, let it pile high on a nearby empty lot, feed it to farm animals, or burn it in their kitchen stove.[3] Development could occur if a farmer set up a route to collect kitchen waste that he could then feed to his animals. As cities grew and land-based resources became scarce, scavengers were likely to approach householders one by one. A board of health might license these scavengers, or even direct them to collect all the garbage from a certain area and require householders to participate. When use of scavengers proved to be inadequate, city leaders might find it easier or cheaper to initiate a municipal collection program. At that point, city officials could decide between setting up the horses, carts, drivers, and collectors on the city payroll or contracting the work out.[4] Once garbage collection programs were up and running, cities continued to grapple with effective methods of collection and disposal. St. Louis built an incinerator, only to have it fail, resorting to feeding city garbage to pigs on an island in the Mississippi. Other cities gave up their high-tech experiments and turned to dumping at the edges of the city. And then the cities had to contend with a population that was slow to alter household practices in keeping with new municipal programs. Residents continued to leave their garbage in open boxes or any receptacle they had. If they mixed ashes with garbage in a wooden box, the mix could become flammable.[5] Even when garbage collection was available, some city dwellers continued to use privy vaults to deposit "an unsightly pile of garbage and refuse of every description."[6] Wooden chutes for wastewater proved a handy vehicle for whisking garbage away from the home, into open drains. "The result was odorous and unhealthy."[7]

As cities struggled with mounting trash, many tried, yet failed, to develop modern sanitary practices. Successful trash collection and disposal requires expertise and technology, which was available but could be expensive. It demands political leaders who are invested in new programs, which many city leaders were not intrinsically motivated to be. It requires an ability to implement new programs. It needs residents to comply, but many resisted (even fought) new trash initiatives that required changing long-held practices. And it draws on political cover when it doesn't work well. For municipal trash collection and disposal to work, residents must follow their city's rules, accommodating their home and consumption to those rules in compliance with a state authority that is nearly invisible. The fact that trash collection today is ubiquitous, that it is viewed as mundane, even nonpolitical, and that residents participate habitually is nothing short of a remarkable government accomplishment.

Nineteenth-century cities across the United States faced a similar problem at roughly the same time. As more and more Americans moved to municipal centers, traditional ways of disposing of trash were no longer sufficient. Trash piled up in city streets, waterways, and yards. To collect trash, municipalities had five needs: (1) Technical expertise and skills: Cities needed to understand the best methods of collection and disposal for their climate. Health officials, sanitarians, engineers, and civic associations had ideas and, at times, resources to address the trash problem. These experts were often sidelined, however, in developing sanitation programs and brought back in during the implementation phase. (2) Political will: Political officials were motivated when they could benefit (politically or financially) from trash collection and disposal programs. Corruption was an important resource to encourage political action. (3) Ability: Cities needed to assemble collectors, drivers, horses, and carts. They needed rules for maintaining these personnel and resources. They needed to either oversee contracts or develop their own administration in public works or departments of health. Corruption could bolster or hinder a city's ability to implement collection and disposal programs. (4) Resident compliance: City leaders quickly learned that formal programs would not work without the support and compliance of city residents. Municipalities relied on gender hierarchy as a resource to encourage compliance. And finally, cities needed (5) political cover to deflect criticism when programs didn't meet expectations. Officials used racial hierarchy, blaming trash collectors and residents for failures of city programs.

Building the American sanitation infrastructure—the technical expertise and skills, political will, ability, resident compliance, and political cover needed to collect and dispose of trash—relied on an unusual combination of means apart from formal, "clean" politics. City governments reached out to available expertise as well as to corruption and to gender and racial hierarchies. Although these resources are not common building blocks in public policy studies, we encountered them as we looked at how municipal sanitation programs were created and maintained and as we examined the tools that governments used when collecting garbage or facing challenges in administration. Following the resources that were used, rather than particular categories or types that we came up with in advance, allowed us to see the range of formal and informal resources employed in governing and the implications for equality and power in the United States.

The early years of municipal garbage collection coincided with the era of progressive reforms, making it all too easy to fold the municipal projects into a wave of "good governing." But many trash programs were not about good governing. They were, at bottom, about maintaining and even growing political power. Political actors used undemocratic means—corruption, gender and racial hierarchies—to accomplish their objectives. When reform waves washed across the United States and the progressives came into government positions, they took the infrastructure developed with these resources and repurposed it for their own ends. The undemocratic elements were not eradicated, however. They remain in policies, reproducing inequality and ascriptive hierarchies. What's most interesting in this story is not what has changed over the past 150 years but how much has stayed the same.

Nineteenth-Century Cities Were a Mess

Late-nineteenth-century American cities were dirty. Soot from industrial factories polluted the air; human, animal, and industrial waste were dumped into rivers and lakes that served as primary sources of drinking water; homes, schools, and factories were crowded and inadequately ventilated; privy vaults overflowed, their contents seeping into soil and wells; urban stables, dairies, and hog farms produced copious amounts of animal waste and flies; and city sewers (where they existed) were clogged by human, animal, industrial, and slaughterhouse waste.[8] It is hardly surprising that city streets were no cleaner.

On the streets of Louisville in the 1880s, one could expect to find trash thrown out by residents, manure deposited by horses, and hogs eating it all up. Trash could pile up to sixteen feet high in Pittsburgh's yards. Household waste and dead dogs littered New Orleans's streets and may have been used as fill to raise the grade of streets. St. Louis floated its waste down the Mississippi River.[9] Cities across the United States at this time faced a mounting garbage problem.[10] Householders had always had ways of disposing of their own waste, of course, but burying trash in the backyard, feeding animal and vegetable waste to chickens or swine, and collecting ashes from fires to use as filler didn't work anymore. When cities were smaller, people could dispose of waste on their own, but as cities grew more populated and people lived closer together, traditional solutions failed. The farmer who may have filled his cart with kitchen waste to feed to his hogs could not remove all of the trash from a large city. Garbage just accumulated. It was hard to ignore the resulting "heaps of garbage, rubbish and manure" that "cluttered alleys and streets, putrefied in open dumps, and tainted the watercourses into which refuse was thrown."[11] Trash created noxious odors, impeded commerce, attracted vermin, and imperiled public health. The problem was daunting, and municipal resources to address it were scarce.

The 1890s wave of garbage ordinances explored a variety of solutions in collection and disposal. Some cities sent municipal carts around to pick up garbage regularly and dumped waste in rivers; some contracted the work out to companies that already had horses, carts, and drivers, and built expensive disposal plants. Others were technically classed as having "no municipal collection," leaving the work up to entrepreneurial scavengers, who went door-to-door in search of trash that could be turned into something more valuable. By 1897 the American Public Health Association's (APHA) Committee on the Disposal of Garbage and Refuse had information on the programs of 149 US cities, whether large, medium, or small.[12] Cities across the country, with varying levels of resources, were joining the movement to provide for garbage removal.

Just because municipal garbage programs were begun does not mean that the garbage actually was picked up. Collection was rocky.

Cities started and failed at experiments in transport and disposal. Householders couldn't or wouldn't play along. And sanitarians— experts in the field—often found themselves watching from the sidelines, their offers to help and inform declined by city governments. Once cities had garbage collection ordinances in place, they actually had to pick up the garbage, which takes resources and capacity. Governments may not have had adequate trucks, wherewithal, or know-how. Trash collection required coordination of horses, carts, and drivers. Disposal meant finding places to dump trash on land, in rivers, lakes, or oceans, or building a plant to burn it. And residents had to participate actively, abandoning the ways in which they had long disposed of trash as they benefited from and complied with new services. The early years of municipal garbage collection offer a study in the development of capacity to do something new: collect garbage from households, citywide, week after week, to maintain sanitation in growing cities.

Nineteenth-Century Garbage Programs

Garbage collection has remained a decidedly local issue. Facing food shortages in World War I, the US Food Administration (USFA) turned its attention to garbage cans. Finding that Americans threw away food that could be fed to hogs or utilized for grease, it erected a Garbage Utilization Division. Garbage became something of "a war-time discovery; something which had no existence in pre-war times."[13] Federal officials who studied garbage disposal found that it varied by city. Researchers compiled information on the volume of garbage collected, the percentage utilized for usable extractions, and methods of collection.[14] This research had been underway for decades, spurred by sanitarians who recognized the decentralized practices of garbage collection and conducted comprehensive studies to identify best practices and make further recommendations.[15] The USFA disbanded at the end of the war, and with it went any federal interest in nationally coordinated garbage policy. Garbage collection remained a local operation.

Local governments are the most common form of government in the United States. They are important not only because policy moves up (e.g., nationalization of education policy) or down the federal ladder (e.g., devolution of social services), but also because local governments have a distinct set of responsibilities that they have come to address. To neglect local governments in studies of politics is to neglect a host of policies that affect Americans directly—from water treatment to street paving to sanitation—and to position politics as something that takes place far removed from their day-to-day lives.

Our approach in this book starts close to home. Governing is an intimate part of Americans' lives, and people see, hear, and feel its effects daily. Sanitation is particularly important in this respect. Nineteenth-century cities tackled issues like clean water, healthy schools, food safety, and air quality.[16] With the adoption of sanitary measures, more people were living longer, especially urban residents. Infectious diseases killed 44 percent of urban residents in 1900, a figure that dropped to 18 percent by 1936. Infant mortality was 140 percent higher in cities than in rural areas.[17] As crucial as it is, sanitation generally, and garbage collection in particular, can be so ubiquitous as to be nearly invisible. More than a century after local garbage collection and disposal programs were created, many residents and researchers have forgotten the politics behind these programs and the fact that garbage programs are basic public health policies.

John N. Collins and Bryan T. Downes, writing about garbage collection and disposal in 1977, felt the need to remind their readers, "Despite frequent humor and snickers about the issue of urban garbage collection, collection and disposal of the mountains of garbage in our cities, or solid waste as it is now called, is a serious business."[18] Even now, when garbage collection and disposal programs are addressed, they are seen as something worthy of a chuckle. Yet as we explain in this book, the creation of garbage programs was often politically contentious, capable of bringing down mayoral administrations. Even when programs were formally created, without capacity, cities failed repeatedly to dispose of trash in a sanitary manner. Residents frequently ignored strict requirements placed on them to sort their garbage and to put it in a specified vessel in a specified location at a specified time. It was no laughing matter. Overflowing, uncovered garbage receptacles and piles of waste created breeding grounds for disease. For a period in which most governing that affected people's lives was done at the state and local level, it makes sense that studies of political development track the mechanisms of change in local governments in seemingly mundane matters. Amy Bridges points out that we can capture a sense of democracy in the experience of citizenship when we do so.[19] A study of garbage collection shows the effect on people's immediate surroundings—what they smelled, where they stepped—their health, and their relationship with political power.

Here we use nineteenth- and early-twentieth-century trash collection and disposal to account for political development in American municipalities. Scholarship on political development—defined as durable shifts in governing authority—tends to emphasize national institutions and policies. Yet political development and state capacity happened much earlier at the state and local level.[20] We examine the shift from purely private action—individual disposal (by burning, burying, dumping in lots, or feeding to pigs)—to programs run or overseen by urban governments. These programs operated in, as Carol Nackenoff and Julie Novkov describe, "locations where boundaries between public and private shifted, where models for state 'borrowing' of private capacity were piloted, where new hybrid institutions were sometimes forged, where a variety of policy entrepreneurs used creative techniques to get results through informal and formal politics, and where institutions and their development can be understood in structural, cultural, and ideological terms."[21] Like many other programs that to contemporary eyes seem at the margins of American politics and policy, late-nineteenth-century sanitation programs were part and parcel of state building by municipal governments across the country.

Together, the individual actions of local governments transformed sanitation in the United States and, by extension, the nation's public health.[22] Yet, as we witness in the twenty-first century, those health gains were unevenly distributed and, rather than a rising tide lifting all boats, generated inequality.[23] In the late nineteenth and early twentieth centuries, American cities began to adopt a "more general concern for residents' wellbeing."[24] To do so, they moved formerly private issues into the public arena, especially sanitary concerns with clean water, clean streets, and garbage removal. As cities across the United States shifted authority for sanitation from individuals to government-organized and/or government-run programs, how did they create these programs? On what resources did they rely? And how have their actions affected the course of American political development, understood as the cumulative action of the nation's local governments and the experience of residents?

In this book we detail how cities developed the capacity for sanitation services, transforming Americans' lives and life expectancies and their relation to the state. We examine the extraordinary resources that they relied on to analyze the garbage problem and come up with possible solutions, to generate the political will to do something about it, to develop the ability to collect and dispose of trash, to enforce citizen compliance, and to deflect blame. Scholars' conceptualization of state capacity has largely focused on the administrative state—the ability of bureaucracies to administer policy through human, budgetary, or institutional resources.[25] Cities established such institutional capacity either by providing services themselves, by contracting out to private organizations, or by doing nothing. Although at first glance it seems that the administrative state building that is required for city collection reflected the most robust capacity, our visits to archives showed just the opposite. Public collection in New Orleans was irregular and haphazard. Private collection in Pittsburgh showed some innovation in garbage collection technology that was admirable for its time. And San Francisco, which technically had no collection, in practice ceded control to a syndicate of immigrant scavengers who provided reliable services.[26] Clearly, formal policies did not equate to administrative capacity, and administrative capacity is not enough to determine what cities chose to do, why, and with what effect.

Dirty Politics

In the early years of garbage collection, cities drew on resources we might consider private, cultural, or inappropriate. The story that emerges is about politics that is messy, contested, and—in this case—exceedingly dirty. Politics is not only following orderly pathways grounded in official channels but also, as we see in garbage collection and disposal, a grab for power. Political actors used resources at hand for political leverage and private gain. Some of the resources are expected (expertise), while others are commonly considered inappropriate and antithetical to the work of government (corruption). Some resources, such as the ascriptive hierarchies (gender, race), seem at first blush to have little to do with governing the disposal of the nation's waste. This book follows the five needs governments have in addressing a pressing public problem and the resources they enlisted to meet those needs. We discuss how local governments met the need for expertise (ignored when inconvenient), political will and ability (corruption), resident compliance (gender hierarchy), and political blame (racial hierarchy). In this introduction we provide a broad overview of the needs and resources we cover in depth.

Available Experts

When garbage was a mounting problem, there was no shortage of expertise. Yet experts were often sidelined by political actors in creating sanitation programs. Sanitarians had been working for years to clean up cities. Physicians found positions in local boards of health, researching and responding to disease epidemics and food safety. They had a professional home in the APHA. At the annual meeting in 1887, a Committee on the Disposal of Waste and Garbage was formed.[27] Two years later, the committee was commissioned to inquire into methods of garbage disposal across the country.[28] Cities had been working on their own methods—whether dumping on land or in waterways, or incinerating, or feeding garbage to hogs. There was no way to standardize garbage collection across the country, and there were no guidelines; cities did what they could with their available resources, given such local conditions as size, population density, and weather.[29] The climate mattered for how often garbage was picked up and even what kinds of receptacles to use: colder climates called for wooden garbage bins, while warmer climates were better off with metal.[30] Some sanitarians were notable in their own right. George Waring's sewerage system was renowned for cleaning up Memphis after a devastating epidemic of yellow fever in 1878. He was likewise famous for his street cleaning program in New York City.[31]

Sanitarians and engineers devoted considerable attention to innovations in garbage collection in the late nineteenth and early twentieth centuries. They conceptualized garbage as a problem for public health. They gathered together in professional associations, collecting data, sharing knowledge, and proposing solutions. Cities had boards of health of one kind or another, which were an avenue for this knowledge, and newly formed public works departments hired for or contracted out giant citywide programs in streets, sewerage, water provision, and garbage collection. These methods appeared in the Progressive Era, a period of social and political reform. Progressives used scientific expertise to improve society. They incorporated social reform into public policy. And they cleaned up governments by railing against corruption.

Sanitarians, engineers, and civic associations were ubiquitous in this era, and they were keen to get involved. We find, however, that these professionals were largely held at arm's length in carrying out garbage collection programs because statements of a problem, and even persuasion that it is a public problem, do not automatically generate a corresponding public response. Cities required the political will to follow through and act upon the sanitarians' concerns. That impetus was provided by another resource available to municipal governments—corruption.

Corruption as Resource

While American cities faced a growing garbage problem, municipal corruption was at its peak. By some estimates, as many as 70 percent of cities during the late nineteenth century had corrupt governments at some point.[32] Corruption is "the abuse of control over the power and resources of government for the purpose of personal or party profit."[33] In practice, corruption took many forms: profiteering through rigged contracting; boodle (bribes) to public officials; and patronage, or giving jobs in exchange for political support. American cities had formal democratic governments at the same time they often had informal systems of power and control: a machine in Pittsburgh, a quasi-aristocracy in Charleston, an oligarchy in St. Louis, the Ring in New Orleans, and a collective of scavengers in San Francisco.

Despite the ethical quandaries invited by corrupt rule, it was only sometimes illegal. In St. Louis, wealthy businessmen bribed city officials to sell them lucrative utilities contracts. Money that could have gone into the city coffers instead lined the pockets of elected leaders and their middleman to the business community. These officials were rarely held accountable. Even when cases came to trial, convictions were rare or were overturned on appeal. In Pittsburgh, much of what the ruling political machine did was inefficient and opportunistic, but it was for the most part legal. The bosses of the machine used their influence in the state and city to write legislation authorizing public works contracts, which were then awarded to machine-connected businesses.

Corruption invited political will, motivating political actors to start trash collection and disposal programs in their cities. Although cities were dirty places, and although trash had begun to accumulate much earlier than the 1890s, city officials were often content to do nothing. Many cities took on trash collection and disposal not when there was an objective need so much as when city officials saw ways they could benefit from it. They wrote contracts to pick up trash that benefited them personally (Pittsburgh); they relied on generations of governing by the same core set of families, taking advantage of racial hierarchy (Charleston); they accepted bribes (St. Louis); and they paid cart owners to collect trash in exchange for political support (New Orleans).

The particular kind of collection and disposal program they chose reflected the corruption endemic in their administrations. In Pittsburgh, the political machine engaged in extensive profiteering, benefiting its allies. City officials created garbage contracts with requirements that *only* the machine-connected businesses could fill. Long after slavery ended, Charleston continued to be ruled by families who, dating back to the antebellum years, had relied on hiring out enslaved people for trash collection and removal. St. Louis's Cinch operated as an oligarchy, creating a ruling class that exercised power through extensive boodling, or bribery. Garbage collection belonged by contract to Cinch businessmen. New Orleans's Ring struggled to retain power in that city's fractured political environment. Ring leaders used garbage collection as a mechanism for patronage, giving important constituencies (in this case vulnerable widows) steady city paychecks in exchange for political support in their wards.

Some forms of corruption were better suited than others for picking up garbage. In St. Louis and New Orleans, corruption motivated political officials to create garbage collection and disposal policies. Yet corruption was at odds with the capacity necessary to carry out successful sanitation programs. St. Louis developed a state-of-the art garbage infrastructure, but it was in the hands of a conduit between political officials and businessmen. The city lost not just the garbage disposal infrastructure but the capacity for modern garbage disposal in a fight to rid St. Louis of corruption. In New Orleans, corruption—extensive patronage without capacity—was at odds with modern sanitary garbage collection, and the city never got a program off the ground.

Surprisingly, corruption both built political will *and* facilitated garbage collection and disposal in other cities. Charleston's governing regime consolidated power and instituted garbage collection in 1806, long before other cities. The existence of slavery in South Carolina proved to be a pernicious sort of resource for the city, which relied on slave labor for collection. Charleston's political officials rewarded the local elites at the same time the city relied on the capacity of enslaved people. The city used the collected garbage as landfill, so it had a need to conduct citywide collection and it used available slave labor to collect it. In Pittsburgh, the machine was materially invested in public works (being the major street paver and owner of the garbage works), so it sought out extensive public projects that would enrich itself. Yet because it had the material resources and business know-how, the machine was actually capable of collection and invested in a state-of-the-art disposal facility. Garbage collection programs developed not despite corruption but because of it.

Resident Compliance

Gender hierarchy, too, was available as a resource of government, but only when local regimes really needed to use it to increase compliance.[34] Women still lacked the right to vote in much of the United States in the 1890s, but they were extraordinarily active in the civic sphere, most notably in the club movement. Both white women and Black women formed their own local clubs, often chapters of national organizations. Women's civic organizations, whether literary, social, or public, were frequently comprised of middle-class women keen to make their neighborhoods and towns—and maybe the neighborhood down the hill—better. They networked with the well-connected men in their social circles. They relied on their personal and social assets to research, gather resources, and carry out civic improvement projects. Cleaning up garbage from the streets, sidewalks, and homes was right up their alley, and well suited to their status, for the doctrine of municipal housekeeping recognized that a woman was wont to clean her own home, go right on out the door, and keep cleaning the area around her until she was engaged in public space.[35]

Despite their eagerness to aid government, women's clubs were seldom taken up on their offers to create garbage collection and disposal programs. Because of the various forms of corruption, governments steered clear of civic organizations. Clubs found themselves left on the sidelines, where they continued to meet and to find sanitation tasks that their city government had not taken on. The Civic Club of Allegheny County—a legacy of Pittsburgh's Women's Health Protective Association—kept track of complaints of expectorating on streetcars. The Civic Club of Charleston provided for public trash cans to be placed on sidewalks. Women kept themselves involved in the cleanliness of public areas of the city. Soon enough, however, when cities ran into a problem getting residents to comply with new garbage ordinances, city governments began to eye local, engaged white women as a resource for implementation.

Despite all the advances in engineering, city officials realized that the garbage simply was not going to get picked up if residents did not put it out, or did not put it out at the right time, or in the right place, or in the proper container. Cities could fine residents for failing to comply, but that took resources, and it was a coercive use of public authority by regimes that were not terribly responsive to democratic norms in the first place. People already had shown that they would flout garbage ordinances. Wielding punitive state power when they did so was only likely to alienate them further. Cities had another resource, however, in "infrastructural power," the capacity of the state to penetrate and use civil society.[36] Infrastructural power makes use of economic and social relations to carry out the needs of the state. The state ensures that social and economic networks operate in keeping with its purposes, and state power is less visible and less intimidating. Encouraging a certain type of citizen behavior in these networks can rest on biopolitics, whereby the bodies and sexual identities of citizens are shaped as well. If citizen behavior and habits are out of keeping with modern needs, the act of shaping involves a seductive lure that makes citizens want to shape themselves.[37] The garbage can problem was well suited for the deployment of infrastructural power. Residents' ingrained habits were out of sync with government goals. State authority needed to reach into the home to induce people to change their behavior. To maintain their legitimacy, city governments had to take a less heavy-handed approach than punitive measures. Here cities found a valuable resource in women's civic organizations.

Civic groups came in handy for taking on the task of compliance by modeling behavior, acting as moral authorities, and even shaming those who kept dirty premises. Quite often civic clubs played up race and class differences in presenting a model householder and stigmatizing those who did not keep a well-ordered household. This ideal status was commodified, with magazine advertisements touting the cleanliness and efficiency of garbage pails and other devices such as the Incinerate, a steel receptacle for burning garbage that was mounted on the wall of a kitchen. The ability to leverage this knowledge in class and race power dynamics is most starkly seen in Louisville and Birmingham, where white women were offered informal supervisory roles in the conduct of Black garbage collectors. When faced with people refusing to comply with public ordinances in their own homes, cities relied on civic activity and cultural power to find ways for public authority to enter into private space.

In doing so, cities enlisted women's civic organizations that had been building their clout on notions of purity. If garbage collection was initially conceived by sanitarians as a mechanism to prevent dangers to public health, civic organizations presented models of purity and clean citizenship that they could wield against poor neighborhoods and people too busy and overburdened to live up to the model.[38] Municipal garbage collection was a measure of government capacity in that cities developed or enlisted new capabilities to offer new services, and it was likewise an opportunity for new growth in state power, requiring new habits in the citizenry and offering new demarcations by which to distinguish citizens.

Deflecting Blame

City collection programs were often poorly run, and when it became evident, city officials often evaded accountability by using racial hierarchy, which already intersected with gender hierarchy.[39] Once city governments started collection programs, there was no one to fault for poor collection other than the city government, so municipal governments found someone else to blame: Black and immigrant Americans. Jim Crow laws allowed states to disenfranchise Black voters, and across the country, employment segregation left many of the most marginalized jobs for people of color. Housing segregation amplified those differences. Immigration patterns saw new groups of immigrants entering cities, bringing new economic needs and new ethnic groups.

Garbage collection is pervaded by racialized difference. Neighborhoods were unequally served by collection schedules set according to the racial and class composition of the neighborhood. And garbage collectors—disproportionately Black or immigrant men—were often held in the same regard as the waste they collected. Carl Zimring describes politicians, businessmen, and reformers who saw the waste trades as "not just physically dirty, but morally degraded. That perception reflected revulsion at the kinds of people who did the work as well as the work itself."[40] By 1920, Black Americans were less than 10 percent of the population of the United States but 27 percent of garbagemen and scavengers.[41] White workers resented Black workers for being strikebreakers in the late nineteenth century. Their labor was normally relegated to the jobs that white workers—including new immigrants—did not want to take, such as teamsters, garbage collectors, janitors, and laundresses.[42] Racial disparities in garbage collection were brought to the attention of a larger public by civil rights groups throughout American history, famously at key moments in civil rights history. The Neighborhood Union of Atlanta, led by Lugenia Burns Hope, highlighted disparities in the way neighborhoods were served with garbage collection, and she worked to build up community aid.[43] W. E. B. Du Bois pointed to the slums of Atlanta as an example of the conditions in which Black families lived in the city, tucked away in alleys with poorly paved sidewalks and streets and inadequate sewerage. Poor sanitation led to the higher mortality rates for Black residents than for white. Du Bois uncovered conditions in St. Louis, too, noting that Black Americans were segregated into a few wards by white flight, landlord discretion, and discriminatory lending practices. These neighborhoods and homes were publicly constructed and then underserved by public works.[44]

Strikes of sanitation workers have repeatedly delivered a profound message because withholding labor points to the invisibility of garbage service as well as the racialized composition of its workers.[45] When garbage workers strike, garbage itself becomes unavoidable as piles of trash mount. Garbage collection then becomes visible as a program with rules and needs. And garbage collectors are rendered visible as workers, operating in a set of working conditions subject to negotiation and possessed of rights. Martin Luther King Jr. was assassinated in April 1968 while visiting Memphis's sanitation workers, who were on a strike that was marked by escalation. On February 1, two workers had been killed while operating a garbage truck, an incident that, according to the sanitation workers' union, reflected decades of dangerous working conditions and low wages. The mayor refused to negotiate, leading to marches, police violence, and larger, more public marches, drawing in wider support. The "I *Am* a Man" placards that strikers carried invoked their basic human dignity while challenging the presumption that sanitation workers were childlike and not fully masculine.[46] Such presumptions lay implicit in the notion of the invisible garbage collector. Other placards made note of working conditions and the need for recognition of the union. Sanitation workers wanted the public to understand that they were public works employees with the same needs of any worker for decent wages, safe working conditions, and a voice.

While such moments have struck the public consciousness in American history, the formal records tend to ignore racial disparities in garbage collection, either in certain neighborhoods or in regard to the treatment of workers. As likely as it was that garbage collectors were Black men, they are hardly mentioned as such in public records in any of the cities we studied. Racial distinctions would have been implicitly understood at the time, but explicit references are elusive in formal documents, with race hardly mentioned in city reports about garbage collection. The infrequency of racial references in the municipal reports can be seen as a small-scale rendition of the "productive absence" of race.[47] That is, just because race is not mentioned does not mean that existing racial categories were not at play. What is evident in the records are periodic hints about the behavior of different groups, whether distinguished by race or by class. The occasional references to race in the complaints we studied tap into "common sense racism."[48] In those few moments when race is invoked, it implies a background of legibility, in that the audience understood what various insults meant. These references often emerge when a group deflects criticism from itself or wields criticism of a regime in order to take over. Racial hierarchy, then, becomes a tool of parties wrangling for control over garbage collection. Like the reliance on gender hierarchy, such instrumental use of status changed that status. These processes are constitutive, on a number of levels.[49] If race is something one "does" rather than something one "has," then race in these complaints is a referent whereby various actors are singled out for what they do or don't do in keeping their residences clean.[50] Reaching into cultural understandings is a process of racialization in which identities are themselves constructed.[51]

Racial and gender hierarchies served as tools for various actors to claim authority. To establish municipal collection, cities often had to wrest control from entrepreneurial scavengers. That transition often involved race. And when political regimes challenged or defended themselves against one another, they could enlist race as well. People of color, who were most likely to be collectors, and poor neighborhoods, which struggled with infrastructural problems under conditions of the extreme inequality of the time, were likely to become the scapegoats for regimes that tried to deflect criticism. There is one episode that flips this pattern in San Francisco, where the Italian immigrant scavengers played up housewives' fears of the Other to maintain their dominance over collection. Other than that, however, racial difference was largely ignored, then invoked—with the attendant stereotypes—when someone got into a spot of trouble.

This use of racial hierarchy was yet another resource of government, folding inequality into political development. By tracking the intersectional operation of gender and racial hierarchies as informal resources of governments, we can see that inequalities do not go away when political development occurs. If a government relies on social hierarchy as a resource of government, then that hierarchy is incorporated into the new, modern way of governing.[52]

Informal Resources of Governance

What can the study of garbage teach us about governing? At first blush, garbage seems both mundane and apolitical. As the vignette at the beginning of this chapter spells out, separating waste from recyclables and taking the trash to the curb are regular, routine habits that many people have developed as a household chore. Yet as we show here, figuring out how to collect and remove trash, how to get residents to comply, and how to deflect blame were significant challenges and overcoming them was an achievement of state power.

Garbage collection illustrates the resources governments draw on to solve ordinary problems and the myriad ways our private lives are intertwined with the instruments of government. To create new public programs, cities drew on formal resources, including public officials and the administrative capacities of existing infrastructure. They also drew on informal resources, including corruption, which incentivized policy officials to address sanitation and to choose particular kinds of solutions that would bolster their political, party, or financial standing; gender hierarchy to induce compliance with garbage programs; and racial hierarchy, which divided cities into spaces where garbage was collected and where it was not, which provided collectors to do the dirty work of picking up trash, and which, when programs failed, served to point the finger at populations and collectors instead of city officials.

Political development does not necessarily mean progress; it can carry over old status categories because governments may rely on these categories to develop new policies and innovations. Old status categories then become embedded in new public programs. In the case of garbage collection, nineteenth-century innovations carry over nineteenth-century biases.

The methodological contribution of this book is to consider the politics of the everyday as intrinsic to our understanding of political development. Our project focuses on garbage, but other researchers could pick other issues—federal, state, or local—map them, and see what happens. What relations does a government foster? What informal resources does it enlist? How does social inequality become a resource for formal governing? And what are the long-term implications when these informal resources are used to build formal public programs?

Overview of Chapters

In this book we tell a story of governance—how cities address local public policy problems—paying particular attention to the resources that they rely on. We look in depth at five key cities—St. Louis, New Orleans, Charleston, Pittsburgh, and San Francisco—and rely on secondary research for two more, Birmingham and Louisville. Our cities are located across the United States, and although the actors and stories are unique to each, we also draw broader lessons from them about how governments take on a new policy program and the resources they rely on to do so.

The story of garbage starts with a counterfactual: If garbage is a big, technical problem, why were experts largely ignored? We show how a political (rather than technical) logic drove decision making around sanitation. We then show how corruption served to generate political will in St. Louis and New Orleans and how corruption provided administrative capacity in Charleston and Pittsburgh. We address how gender hierarchy was employed to generate compliance while racial hierarchy served to deflect blame.

We start in chapter 1 with a conceptual roadmap that lays out our theoretical and methodological approach. We trace how governments achieve their goals, rather than limit our analysis to preconceived governmental actors and institutions. An academic audience interested in the foundations on which we build our analysis can begin here. But an audience interested instead in the story of garbage collection and disposal can easily skip this chapter and move right on to chapter 2.

Chapter 2 presents the garbage problem that cities faced as they grew and became more concentrated. As the problem arose, so did possible solutions, with technological advances that experts recommended. Sanitarians and engineers devoted considerable attention to innovations in garbage collection in the late nineteenth and early twentieth centuries. They gathered together in professional associations, collecting data, sharing knowledge, and proposing solutions. Yet even after all of the research and organizational developments made by these professionals, they were shunted aside in city politics. Cities had boards of health of one kind or another, which were an avenue for this knowledge, but they were continually beset by the politically connected public works departments. Boards of health offered the pretext of concern for public health, while public health officers themselves were largely marginalized in the political game. The identification of a problem and knowledge of viable solutions can be insufficient to bring about change to or with governments. Sanitarians and reformers studied garbage collection, but they needed a route to the political process. This chapter tracks the rise of professional associations, civic organizations, and the networks connecting them. The establishment of knowledge allowed for the passage of garbage ordinances, but these experts did not control municipal collection and disposal decisions or programs.

Experts raised attention to garbage as a political health problem, but they did not generate the political will to do anything about it. Instead, we show how corruption motivated city officials to address garbage collection and disposal. Chapter 3 shows how corruption generated political will but failed to create lasting solutions in St. Louis and New Orleans. Corrupt regimes were willing to take on collection and relied on their various connections to profit from it. Corruption, then, was a resource that many governments chose over available expertise to respond to the garbage problem. As expected, corruption did not make for good government. St. Louis developed the capacity to collect and dispose of trash, but the city lost its ability to do so in a bid to rid the city of corruption, while New Orleans's disposal plant never got off the ground. What is it about corruption that makes development so hard? This chapter highlights what many people already believe: corruption makes good policy harder.

Chapter 4 shows that corruption *promoted* policy development in Charleston and Pittsburgh. Although the form of corruption endemic to each city differed, in both cases it generated political will and political capacity to collect and dispose of trash. This chapter shows how corruption can actually encourage governments to adopt progressive policies and innovative technologies when they stand to benefit. It highlights what is counterintuitive to what many people believe: corruption can produce policy development.

Once cities established their methods of collecting garbage, they faced another problem: getting householders to actually make their waste available to collectors. The sanitarian Charles Chapin called this "one of the most commonest [sic] forms of nuisance to be found."[53] Once policies are put in place, governments need compliance from householders. How do governments get people to change their habits and behavior to make garbage collection successful? Chapter 5 identifies white women's civic organizations as the available resource. This chapter shows that municipal governments, which largely ignored women in creating sanitation policies, enlisted some formerly sidelined women's groups, which had been eager to get involved in garbage collection in its early years. Engineers carved out new opportunities and new management of projects because of the growth of municipal public works, including garbage disposal. Women found new opportunities brought about by the garbage can problem. In addition to covering the women's groups in our primary case cities, this chapter also looks at the opportunities that women in the southern cities of Birmingham and Louisville found for themselves. Those cities gave women formal opportunities to supervise employees. The power dynamics in those cases allow us to see what ensued in our primary case studies, cities where women lacked access to formal institutions but carved out roles and authority in civic spaces. Women eagerly offered their services as resources of governing, building on the intersection of race and class privilege and gender hierarchy.

Governments also need collectors to pick up trash and dispose of it effectively, and people on whom to deflect blame for poor policies. What resources did they use? Chapter 6 discusses the long history of sanitation's relationship to racism, fear, and deflection, as cities drew attention away from their own shortcomings and turned it toward specific city populations. This chapter traces the history of racializing garbage collection back to when cities initially wrested collection practices away from individual scavengers. Once garbage collection programs were run out of departments of public works or health, problems of collection could be attributed to laggard citizen populations rather than the reigning political regime. Scholars have demonstrated that racial hierarchy played a role in both who was employed as garbage collectors and which neighborhoods were ignored for collection. This chapter shows how racial hierarchy also was used as a resource of government to blame collectors and poor neighborhoods for inadequate trash collection and disposal programs. The chapter takes up the unusual garbage collection program of San Francisco, categorized by sanitarians as having no municipal collection. In place of a city program, San Francisco relied on scavengers who made arrangements directly with householders; it was a reliable program, and the most enduring. While this outlier did not make San Francisco a study of corruption, the scavenger-householder relation offers a stark depiction of the ways in which racial innuendo can be enlisted in maintaining power.

Chapter 7 considers lessons learned about state capacity, political development, and gender and racial politics. As a study in political development, this book shows how innovations in technology and state capacity usher in new relations between a state and its residents. Viewing governance through the resources that the state enlists to get its work done illuminates not just formal resources but those informal, unseemly resources too. We find that factors such as corruption, gender and racial hierarchies, and the intersection of the two are enlisted when governments tackle public problems. The use of these factors explains why old ideas get incorporated into modernized regimes and state capacity and how they last over time. Governments need resources, and they may rely on social hierarchies, not necessarily to reproduce them, but because they are available and useful as a resource to achieve other ends.

The conclusion emphasizes the importance of politics not just as what happens in Washington, DC, and electoral participation. Even the most mundane issues—like trash—illustrate how politics and power operate all around us, often hiding in plain sight.

Today, we take garbage collection and disposal largely for granted. But it was a struggle to put sanitation programs in place and to get people to comply. Understanding the resources that governments use to create and maintain policies gives us a more thorough depiction of the way governments make and maintain policy and how political development occurs. It also shows how legacies of inequality are built into the most mundane and ordinary practices, like garbage collection and disposal, to meet the imperatives of governing in a developing public program.

CHAPTER 1

A Conceptual Roadmap

Theory and Methods

Researchers have identified a variety of resources governments use when they take on a new policy, from money to electoral victory margins to reputation. Our previous research on the role of family in law and policy suggests that governments use an even wider range of resources to achieve their goals.[1] Sometimes those resources are familiar and well established in the literature (money, party loyalty, and agency reputation),[2] and sometimes they are unfamiliar and are only emerging in the literature (seemingly unrelated organizations, private relationships, and other levels of government).[3] To discover how governments accomplish their objectives, we look at a prominent problem in the United States (trash in the nineteenth century) and ask: How did governments address the garbage problem? Why? And what resources did they employ?

Our goal is to understand how governments work by examining them *in relation to* other actors, institutions, and levels of government. We trace government actors and institutions to the host of governmental and nongovernmental entities that were important in defining a problem, creating policy to address it, and implementing solutions. The approach we use in this book mirrors a broader approach in both public administration and political science, moving from government (defined actors and institutions) to governance (collaboration with a broader array of third parties to achieve policy objectives).[4] We focus less on particular mechanisms, however, and more on relationships and resources.

In this chapter we describe the nontraditional resources that proved important in the story of nineteenth-century trash collection and disposal using the governance approach. We discuss how American political development (APD) research serves as a foundation for our own. We detail our methodology and provide a preview of the lessons we learned through this study. (Chapter 7 situates our findings in broader theoretical work.)

Governance and American Political Development

When cities attempted to collect trash, they did not limit their efforts to elected officials and public agencies. Leaders reached outside formal government, to the tools available in the economy and in society, what public administration scholars and others call governance. Scholars across disciplines have acknowledged the broader range of resources used to carry out government functions.[5] Carol Nackenoff and Julie Novkov illustrate how private resources are used for public ends. The concept of governance allows us to see not just the state but the state in relation to other actors, institutions, and resources.[6] As Jon Pierre and B. Guy Peters explain, the focus "is not so much about structures but more about interactions among structures. We should expect governance to be dynamic with regard to both configuration and objectives: the inclusion and influence of different actors could well change over time and across sectors."[7] Focusing on governance allows us to unearth the formal means that officials rely on, such as offices and ordinances, as well as those informal (and even undemocratic) means that are employed, such as corruption as well as gender and racial hierarchies.

When we abandon preconceived ideas about who is a political actor and what tools governments use to achieve their objectives, when we follow the many different individuals who pass in and out of the story and examine what they do, how they do it, and with what effect, we are studying governance. In this book, we tie the actors, the diverse tools they use to accomplish their objectives, and the effects together to better understand how governments create and manage new public services. Governance includes formal actors (mayors, legislators, ward leaders, agency heads) and formal resources (infrastructure, revenue, expertise), informal extra-governmental resources (corruption, gender and racial hierarchies), and their interactions with one another. When formal government services employ the social status of race, class, and gender, this shapes both the services that cities provide and the ascriptive identities of the citizens they serve. Citizens enlisted to carry out public purposes can find their status elevated or denigrated, and they in turn can shape the state itself.[8]

Governments needed resources to collect garbage (or make it look like they did). In those instances, they reached out past the boundaries of the formal state. We find that cities relied on race, class, and gender hierarchies as resources in garbage collection. The ideas embedded in racial, gender, and class inequality enter administration not through hostile discriminatory actions but through paternalistic efforts to live up to the public purposes of sanitation. In this way, ascriptive status can be carried into a new institutional order. Desmond King and Rogers Smith have attributed the invocation of race to policy alliances and political imperatives.[9] We identify it through the imperatives of governing.

Because we ask how governments achieved garbage collection and disposal, these hierarchies aid municipal governments in addressing the problem of capacity. They are not immediately apparent as such. Contending with the weak American state, scholars have identified the operation of the hidden state and the decentralized state, which explains not only the carrying out of public purposes by informal actors, but also the disparate effect of public policies when the state borrows capacity from informal or local actors.[10] The coercive capacity of the federal state was already in place to control and shape the status of marginalized groups. As Chloe Thurston observes, "State power is wielded in ways that have different effects on different groups."[11] Similarly to scholars who have demonstrated how the federal state operates in nontraditional ways, we find that municipal state building relied on a hybrid of public and private tools, which included those factors we frequently think about (subject matter expertise, formal agencies) as well as those unsavory and undemocratic factors we don't often think of as resources of governing (corruption, gender and racial hierarchies). Both the accepted, formal democratic resources and the informal, undemocratic resources became part and parcel of municipal sanitation capacity and, by extension, American political development. Although we examine in this book four key resources (expertise, corruption, and racial and gender hierarchies), we explain in what follows how the three undemocratic ones—corruption and racial and gender hierarchies—are part and parcel of governance.

Corruption

Corruption is often defined as "behavior which deviates from the formal duties of a public role because of private-regarding (personal, close family, private clique) pecuniary or status gains; or violates rules against the exercise of certain types of private-regarding influence."[12] We favor V. O. Key's definition of corruption as "the abuse of control over the power and resources of government for the purpose of personal or party profit" because it connects corruption to power and politics, not just illegal activity.[13] In fact, under Key's definition, corruption can be either illegal or legal.[14] Scholars of American politics have examined corruption in American cities in the same time period that we study. They have shown the power and limitations of machine governments.[15] Yet for all of the attention to nineteenth-century political machines and electoral behavior, there is also a "supply side" to machines, in that they increased the public sector in order to have more jobs to give out to their voters.[16] And municipal reformers were capable of producing city governments that were exclusionary and weakly responsive to constituents.[17] Corruption can be considered in American politics more broadly, especially corruption as a resource of governing.[18]

Corruption can create political will to address public problems, like trash collection. Corruption can also provide the capacity to do so effectively. In prior work with Elizabeth Pérez-Chiqués, we show how the logic of corruption determined municipal garbage strategies.[19] When cities chose to take on garbage collection, the kind of collection and disposal programs they chose were intimately related to how the ruling political regime would benefit.

Nineteenth-century cities were often ruled by corrupt regimes. But the type of corrupt regime and the relationship between corruption and democratic institutions varied across cities. In short, corruption looked very different across the United States. Table 1.1 lays out (1) the type of corrupt regime, whether machine, oligarchy, aristocracy, or ring; (2) the mechanisms these regimes used to exert influence and power, including rigged contracts, boodle (bribery), nepotism, and patronage; and (3) the strategies these corrupt regimes used to exploit garbage collection and disposal programs for their political or financial benefit.

Table 1.1 How the logic of corruption determined municipal garbage strategies

City	Type of corrupt regime (mechanisms)	Garbage strategy	Money flows from/to	How does it relate to power?
Pittsburgh	Machine (rigged contracts)	Rig city contracts for disposal plant/collection to ensure machine businesses get contracts	City coffers → machine-controlled businesses	Enrich machine, consolidate its political power
St. Louis	Oligarchy (boodle)	Bribe city officials to ensure oligarchy (Cinch) businessmen get disposal plant and collection contracts	Corrupt business-men → corrupt politicians; city coffers → corrupt businessmen	Enrich business-men, politicians for hire
Charleston	Aristocracy (nepotism)	Elite governing families hire out slaves of city's elite to collect trash	City coffers → Charleston's elite	Reinforce political and financial advantages of elites
New Orleans	Ring (patronage)	Weak democratic factions used patronage for women to shore up political control	City coffers → working-class ethnic constituencies	Bolster electoral support, political power
Columbus	Progressive	Follow expert advice	City coffers → engineering and technical staff	Providing good garbage services bolsters electoral support and political power

Source: Patricia Strach, Kathleen Sullivan, and Elizabeth Pérez-Chiqués, "The Garbage Problem: Corruption, Innovation, and Capacity in Four American Cities, 1890–1940," *Studies in American Political Development* 33, no. 2 (2019): 6. Reproduced with permission.

Compared to the benefits of good, clean government, corruption comes up short. It takes resources that could be used to promote the public good and instead pads private accounts. We are not advocating a return to corrupt governments. The reality in the nineteenth-century United States, however, was that many governments were corrupt, and corruption served as a resource to generate political will for local officials to do something about the growing trash problem when they saw a way they could benefit. Of course, in some cities, corruption also hindered governments, making it impossible to provide good services. In other cities, though, corruption provided the capacity needed to carry out successful garbage collection and disposal.

Gender and Racial Hierarchies

Gender and racial hierarchies too served as a resource for nineteenth-century governing around trash. Scholars have detailed the inequalities that women and people of color faced in providing input into both government and the types of treatment and services received from government.[20] We draw on and incorporate this research into our own. The governance approach, however, also shows how gender and racial hierarchies are employed as available resources to ensure compliance and deflect blame when policies on paper do not work in practice.

Certainly, scholars have shown the importance of gender in nineteenth-century government. Women were left out of formal politics and used civic organizations to exert influence in local communities.[21] In studying trash collection and disposal programs, we found that women played an important role. They had very little authority in creating the programs, but governments turned to women when they needed help with house-holder compliance in implementing them. Some cities hired women to go door-to-door. Other cities held white women up as model householders, providing a standard for others to emulate.

A body of research shows that race structured who collected trash (disproportionately Black and immigrant men) and which neighborhoods were well served or ignored.[22] While we build on this research, we also show how racial hierarchies were used—repeatedly, across cities—to deflect blame for poor services. In short, racially diverse neighborhoods, often neglected in trash collection, were *blamed* for their sorry state and the poor service they received. As a resource, racial hierarchy provided a means for laggard regimes to avoid criticism, while reshaping the status of citizens on the basis of race and class.

Local American Political Development

A study of the early years of nationwide municipal garbage collection offers a somewhat novel local perspective on American political development. While political development can be captured most broadly by durable shifts in governing authority, the field tends to focus on the growth of federal authority, exercised through the administrative state.[23] Municipal garbage collection nevertheless has the trappings of the more familiar modes of federal state building. Local governments gained their own authority and capacity to carry out new government functions. The state is present and developed, with new departments and experts occupying city offices, overseeing a labor force and the machinery of collection, or contracting out to such experts. The addition of these agencies is durable. Much as it did in the 1890s, garbage collection continues to operate at the local level, with trucks picking up cans set out by households, in accordance with that city's schedule. Garbage collection rested on the incorporation of emerging ideas about sanitation, public health, technological developments, and municipal reform in a time of growing urban populations whose concentration presented immediate physical challenges to the old ways of disposing of waste. There was a nationwide wave of ordinances in the 1890s, coordinated by sanitarian and municipal reformers, who shared knowledge and political resources. Cities established new capacity to carry out the new garbage ordinances, whether by hiring city employees housed in a city agency or contracting the work out to a private collector.

Examining local development is not merely a matter of smaller scale; it offers the opportunity for comparative study, upending some of the presumptions we bring to bear about what development requires and what it means for a government to develop. We started by focusing on public decision making around garbage collection and disposal, tracing how formal political actors chose to address the garbage problem, how they determined the solutions they would use, and on what actors or relationships they relied.

Development implies a reaching from what has gone before toward new, better, more effective mechanisms.[24] But that was not borne out by the details of city experiences. New Orleans, with city collection, promised that it had developed administrative capacity, which we found not to be the case. Pittsburgh's contracting out of collection and disposal services or San Francisco's categorization of no municipal collection seemed to indicate a lack of capacity, yet those cities proved to be somewhat innovative and capable. This study of political development likewise leads to encounters with a seedier side of politics, which we expected would get left behind as cities developed. Yet municipal records indicated that corruption as well as gender and racial hierarchies accompanied the development of garbage collection. In fact, some of those unseemly resources produced the best garbage collection. Pittsburgh's entrenched Republican political machine, with a contract given to the machine leader's brother, actually picked up the garbage, invested in new disposal technology, and persisted until World War II. The new Democratic regime even relied on it. San Francisco, classified as having no municipal collection, depended on a tight group of Italian immigrants who made arrangements with individual households. The legacy of that scavengers' union continues to operate in San Francisco to this day. Clearly, public collection itself, with its generation of state capacity, is not a marker of development. And corruption is not necessarily an impediment to development. The comparative study of cities suggests that development depends on the political context of a particular city.

Local context can make sense of these features. American political development scholarship recognizes that change is always built upon existing institutional arrangements, what Karen Orren and Stephen Skowronek describe as "prior political ground."[25] The vestiges of old orders accompany political development. New practices, rules, leaders, and ideas are layered on top of old, creating disjunctures in institutional arrangements. Cities are particularly well suited to studying development because cities have many opportunities for disjunctures—within cities, among cities, and/or between cities and states—which can generate friction and bring about change.[26] Decisions about what method of garbage collection to use largely rested on the vagaries of the reigning political regime.

Taking into account the practices, rules, leaders, and ideas that make up prior political ground in the development of municipal garbage collection, ideas play a particularly important role in our story. Ideas are the foundations on which institutions are built, and ideas continue to affect development after the institutions are in place. Ideas are not just the vehicle for contradictory purposes; once change happens, they can act as an "emulsifier" to frame a new world in which institutional development can then take shape.[27]

The power of ideas to shape political development does not mean that ideas are self-enacting. Our case vividly illustrates how an issue can be both a legitimate material problem (reeking garbage accumulating in city streets) and an idea (government has authority to do something about mounting garbage as a public health problem). Nevertheless, merely because there is a material problem does not mean that anything will be done about it. John Kingdon famously asked, "How does an idea's time come?"[28] In the late nineteenth century, garbage had to be crafted in ideational terms as a problem requiring a public response. Sanitarians and public health experts pitched it as a public health problem, an approach that took hold. Cities across the country passed ordinances to address it, and many corrupt regimes were quite capable of simply carrying out collection themselves. In fact, the more power they had, the better job they did. Corrupt regimes were adept at taking a reformist idea and not ceding its administration to reformers. They could claim to be carrying out the public purpose of sanitation, and to make good, they actually had to get it done.

Importantly, ideas can be co-opted as well, and ideas put forward for one purpose can be used for another. Jessica Trounstine shows that machine and reformist regimes were not all that different; each wanted to consolidate its power, and both could do so by providing similar services.[29] Machines and reform regimes alike could tap into seemingly reformist ideas and retain power to serve their own interests. Corrupt regimes—like the ones in this book—take up popular reformist ideas to secure their own power.

Methodology

To understand how governments across the country dealt with the growing garbage problem, we turned to the indefatigable data collection of contemporaneous sanitarians who classified cities as having public collection, contracting out, or having no municipal collection. We collected archival documents starting with the papers of city officials and city agencies, expanding from there to relevant civic associations and individuals. At times our work was straightforward. Mayors' annual reports included records from each city department, cataloging their budgets, their accomplishments, and their frustrations. Archival records show political actors and officials who may never have imagined their papers would be made public or who may not have cared. We found copies of letters from the head of Pittsburgh's political machine detailing his day-to-day business practices, as well as court cases and newspaper stories depicting the machine's structure and influence. As Pittsburgh's city controller once noted, corruption was not an open secret in Pittsburgh; it was not a secret.[30] But at other times, the relationship between formal government decision makers and the resources they relied on was murky. In New Orleans, the city gave a garbage collection and disposal contract to Maurice Hart, who is described by reformers as an unsavory character. But Hart's relationship with the formal institutions of government were hinted at but never spelled out in any detail.

Newspaper accounts and signs of partisan tension in formal reports indicate that there were political scandals, competing regimes, and selective information reflected in the gaps in public reports. Once ordinances were passed, cities actually needed to pick up garbage, relying on whatever public or private means were available to meet the required capacity. City officials created trash collection programs that depended on resources far outside what was ethical and, many times, legal. They wrote contracts to pick up trash that benefited them personally (Pittsburgh), accepted bribes (St. Louis), and paid cart owners to collect trash in exchange for political support (New Orleans). They drew on resources that were undemocratic: they used racialized gender hierarchies to promote model behavior for householders and to shame residents into following new procedures (Birmingham, Louisville), and they used racial hierarchies to shift the blame in a number of our cities when garbage collection and disposal programs—dictated by politics rather than evidence—inevitably fell short.

In this book we tell the story of how late-nineteenth- and early-twentieth-century cities developed garbage collection and disposal through a comparative municipal study drawing on deep-dive archival research into five cities—San Francisco, St. Louis, New Orleans, Charleston, and Pittsburgh—and, to balance our findings, secondary research on two additional cities: Birmingham and Louisville. As Raul Pacheco-Vega notes, comparing similar-sized US cities "facilitates contrasts between relatively different units . . . within the same national context."[31] It also avoids a "shallow comparison" that can occur from looking at only reported figures, federal or local.[32] We chose to steer clear of the largest, most corrupt American cities with detailed histories written about their local governments—New York, Chicago, Boston—because we wanted to focus not on the exceptional nature of the city or corruption in government but on the *commonalities* that drive cities to take action and the particular solutions they chose.

Our five deep-dive cities were major municipalities of similar size spread across the various geographical regions of the country (West, Midwest, South, North).[33] They all had some corruption during the time period of our study (1890–1929).[34] Because there are no "typical" southern cities during Reconstruction and the Gilded Age, and because the particular forms of corruption and garbage collection, such as the curious use of "widows' carts" in New Orleans, were unlike what we were seeing in other places, we oversampled southern cities. At the time, western cities were few and small. San Francisco, the biggest West Coast city, had no formal garbage collection, although that categorization hides the fact that garbage actually was collected, by independent scavengers who contracted directly with city residents.

From these archival visits, we collected thousands of pages of data from city archives from the 1880s until as late as 1940, drawing extensively from the cities' annual reports generally, as well as the departments of health and public works in particular. We also collected nineteenth-century newspaper articles and trade magazines on sanitation, garbage, and the major public players in our stories.

The details of the cities themselves—the colorful key players and the actions they took—were unique, and we present context for four cities (St. Louis, New Orleans, Pittsburgh, and Charleston) in the chapters on corruption (chapters 3 and 4). But the book overall examines the patterns that we found across cities. Late-nineteenth-century cities relied on similar resources to develop municipal garbage collection programs: experts played a diminished role in creating public programs (chapter 2); corruption was important for generating the political will to take action on garbage (chapter 3) and providing the capacity to pick up trash (chapter 4); cities relied on gender hierarchy to encourage and enforce compliance (chapter 5); and they employed racial hierarchy to deflect political blame for poorly operated city programs (chapter 6).

To ensure that the story we were sketching was not an artifact of the particular cities we had chosen, we collected a small number of primary archival documents and/or secondary data on three additional cities: Birmingham and Louisville, which appear explicitly in this book, and Columbus, which is an implicit comparison. Again, the choice of these cities added confidence that we had found durable patterns, even if the names and the players changed.

Contribution

Our study of the development of municipal trash collection and disposal programs offers lessons for scholars of politics, policy, and history (see chapter 7). First, our book shows how policy problems and contentious disagreements—both between political actors and between political actors and the residents they represent—grow to be nonpolitical over time in ways that hide government power. Because municipal residents often see trash collection as mundane and nonpolitical, they don't think about the ways they are carrying out governmental objectives through their daily habits, while scholars don't always consider what it takes to habituate residential behavior.

Second, our book shows the resources that governments rely on to achieve their objectives, including both formal government actors, institutions, and relationships and informal actors, institutions, and relationships. Mapping what these formal and informal resources are and how they are related to one another gives a more accurate (albeit messier) picture of how government works. To create the sanitation infrastructure on which Americans now depend, government relied on unsavory and undemocratic elements.

Third, political development is not the story of the progressive forward march of liberal values. It also includes incorporation of these unsavory and undemocratic elements into the very fabric of public policy. One striking feature of contemporary garbage collection and disposal programs is how much they resemble the original programs put in place in the late nineteenth century and how little has changed. It's a matter of not only similar collection and disposal technologies but also similar hierarchies in terms of how practices are modeled and who is blamed when programs fail.

The following five chapters sketch the story of garbage collection and disposal, providing detail about the specific cities, the corrupt regimes that ruled them, and the resources they used to address the garbage problem: expertise (chapter 2), corruption (chapters 3 and 4), as well as gender and racial hierarchies (chapters 5 and 6). Chapter 7 situates the findings of this book in the broader academic literature, while the conclusion adds contemporary context.

CHAPTER 2

Ready to Help

Experts Urge Municipal Garbage Collection

American cities saw rapid growth in the late nineteenth century. Between 1870 and 1890, Pittsburgh's population increased from 86,076 to 238,617, New Orleans's population from 191,418 to 242,039, St. Louis's from 310,864 to 451,770, and San Francisco's from 149,473 to 298,997.[1] With increase in size came a greater concentration in living conditions, leaving residents with less space to dispose of their trash.

Unsurprisingly, these cities were a mess. They were faced with human waste, household waste, commercial waste, as well as dogs defecating and people spitting in the street. The sanitary engineer George Soper observed that the city street was "a natural outlet for filth." Householders swept dirt from houses onto the sidewalks. Garbage left out could be scattered.[2] If householders kept their garbage on their lots, they might stow it in the space underneath their homes, on soft and boggy ground.[3] Or they might just pile their garbage on empty lots. One lot was reported to have a pile of garbage sixteen feet deep.[4] Vacant lots might also be used for dumping from night scavengers (who emptied privies).[5] Garbage might get dumped into the river, but then it had to be hauled away lest it pollute the riverfront.[6]

Residents had traditionally rid their homes of waste on their own, but the accumulation of these efforts led to mounting piles of trash. With individual solutions no longer tenable, garbage was becoming a public problem requiring a public solution.[7] Cities had to contend with trash if they acknowledged it to be a public problem. Residents in this period certainly put up with plenty of odors and unpleasant conditions, but these were not always recognized as problems in need of solving. As factories increasingly appeared in urban centers, the smoke they emitted was even considered a sign of prosperity and industry. "Pollution" was not a word in the American vernacular until the late nineteenth century. To trigger a government response, garbage needed to be constructed as a social problem and as an issue requiring a public response.[8]

There were plenty of experts available with information about the problem and potential solutions. By the 1890s, there was a network of sanitarians, engineers, municipal reformers, and civic organizations, each group offering its own specialized knowledge and resources and combining their effort with one another. They were organized into associations, whether civic or professional. They produced their own trade publications, hosted their own national meetings, and collaborated with fellow organizations. They identified sanitary problems, envisioned solutions, and proposed policies that would take into account both public health and administrative capacity. They targeted garbage as one of many issues, along with sewage, water provision, street cleaning, and lighting. The rise of experts in public health, municipal reform, and technology, as well as the interests of various civic associations, offered knowledge and capacity.

Cities responded with a wave of garbage ordinances passed across the country in the mid-1890s. The initiative can be attributed to those reformers and experts. The actual operation of those ordinances, however, largely sidestepped the reformers, who were mostly responsible for introducing them. Offers to help would be rebuffed or underutilized by corrupt governments, which were happy to put forward the garbage collection ordinances but also happy to carry them out themselves.

The unexpected lapse between reformers' ideas and their participation can be explained by politics rather than by best practices. By "politics" we mean both the formal, official policies and procedures and an informal set of rules and practices that marshals whatever resources it can (corruption, inequality, and undemocratic actions). This chapter traces the rise of various types of experts with plenty to contribute to the defining and solving of growing urban problems, while also marking the points at which available expertise was pushed aside.

Sanitarians

Sanitarians, medical experts concerned with public health, were among the first to define garbage as a problem. Sanitarians initially made a connection between a person's surroundings and health, promoting the "filth theory of disease." As members of a movement, sanitarians sought to change public knowledge, habits, and culture in order to realize this connection.[9] As a political interest, sanitarians worked to change laws and policies to get governments—federal, state, and local—to develop administrative capacity and invest in infrastructure that would promote health by making homes, streets, pipes underground, and the air above cleaner.

Originating in England, the sanitary movement took hold in the United States with the professionalization of physicians in the founding of the American Medical Association (AMA) in 1847. The AMA was instrumental in gathering vital statistics of births and deaths.[10] The Civil War boosted the profile of the sanitary movement with the operation of the Sanitary Commission, a civilian organization authorized by the federal government to provide medical and sanitary assistance to Union soldiers. The Sanitary Commission was not enduring, but the visibility of its efforts promoted the movement. The sanitary movement was available to address the dirty streets and piles of garbage that filled cities as they grew in the coming decades, as well as questions of how to provide clean water to cities and how to flush out sewage. Sanitarians were concerned with filth in urban surroundings and the devastation of epidemics, as yellow fever and cholera hit the concentrated populations of cities.[11]

Combating epidemics required sanitary infrastructure, so sanitarians turned their attention to urban planning. Sanitarians organized themselves into a professional association, founding the American Public Health Association (APHA) in 1872. Through meticulous study and data collection, they conducted surveys of cities which identified every street, lot, and building to home in on the concentration of disease.[12] The first meeting of the APHA convened medical professionals on boards of health in various cities for "the advancement of sanitary science and the promotion of organizations and measures for the practical application of public hygiene."[13]

In light of this mounting interest in public health and the interstate nature of disease, the federal government would seem the likely place for a government response. Indeed, a national board of health was founded in 1879, in response to Memphis's devastating yellow fever outbreak. The board could have centralized knowledge about and responses to the garbage problem, but the federal organization was too intrusive. Business interests (which already resisted quarantines) decried the board as "coercive and restrictive of trade," and it was disbanded by 1883.[14]

Even though national sanitarians might know a great deal about the garbage problem, it was determined that the federal level was not the most appropriate site to implement a sanitary program of garbage collection because solutions needed to take local context into account. The APHA formed a committee on garbage disposal in 1887. It studied available methods and recommended reduction, an innovative method that extracted materials from the processing of garbage, and incineration as the most promising approaches. By 1894, the committee had collected "elaborate statistics" from more than one hundred cities.[15] The APHA committee found that there was no single preferred method, as the best choice depended on local conditions.[16] Such conditions might be as disparate as weather, funding, or political will. Sanitarians determined that southern states had special needs, not just because of their warm climate but because southern households consumed more fruits and vegetables than northern ones, so they produced more garbage per capita. Northern states used more coal for heating, which generated more ash in household garbage.[17] Those considerations would determine the amount of waste a garbage can would have to accommodate, the kind of can to use, whether garbage needed to be separated for the collectors, the frequency of collection, and the method of disposal. The local refrain was repeated in sanitarians' studies of garbage collection. Determining the proper method for a city would require "preliminary investigations of the conditions prevailing in each locality."[18]

Boards of Health

The sanitary movement found an institutional home in local boards of health, which dated sporadically back to the early republic. Although yellow fever was common in warmer southern climates, it broke out across the country as port cities—in the North and South—were hit. Philadelphia's 1793 outbreak paralyzed the city, and nearby towns rushed in with supplies and reinforcements. Such epidemics were repeated so frequently that 1793–1806 can be labeled the "yellow fever era."

Cities generated particular responses. Philadelphia relied on voluntary organizations to supply the services needed to weather the epidemic. New York City, by contrast, established a health committee, which included two doctors, with the authority to impose quarantines, under the theory that ships were importing the disease. Cities tended to erect a board of health as a reaction to a public health scare and then dismantle it when the danger passed. Charleston saw the early institution of a board of health in 1815 because of its climate and port activity. Both a port city and a wet, hot, southern city, New Orleans had a critical need for health authorities, but it lacked the will to maintain them. New Orleans had a board of health as early as 1804, albeit "in name only."[19] It soon disbanded. In 1818 the city established a new board of health, with physicians as officers, who would enact quarantines. It was repealed the following year. A third board of health lasted from 1821 until 1825.[20] Despite the frequent and serious episodes of yellow fever in the antebellum era, the board of health was abandoned after each outbreak. In 1852, for example, after a swamp-draining program was put in place and yellow fever was successfully eliminated for five years, the board's members were let go, with only a secretary remaining.[21] A board of health that now existed in name only faced a yellow fever epidemic the next year.

Boards of health laid the groundwork for municipal commitment to public health. They had authority dating back to the Anglo-American common law heritage, largely through nuisance abatement. If water from a neighbor's roof drains onto the house next door, the next-door neighbor can claim a nuisance. If the smell from a neighbor's pig farm wafts into the yard next door to impede enjoyment of property, the neighbor's pigs are a nuisance. As the English jurist Sir William Blackstone explained, "any thing that worketh hurt, inconvenience, or damage" is such an annoyance and, in the common law tradition, qualifies as a nuisance.[22] If a person blocked off a road and impeded access more generally, then that nuisance was a public one. Regardless of the type of nuisance, the common law recognized that one person's activities could have an effect on another person. By the early twentieth century, sanitarians identified nuisances in filth on private property, defective plumbing and drainage, cellars, overcrowded housing, weeds, poison ivy, streetcars, noise, spitting, stables, manure receptacles, privy vaults, public dumps, and offensive trades such as slaughterhouses.[23]

Under common law tradition, nuisances were taken care of in courts, which could issue a fine and an injunction to cease the activity. A judge drew upon the authority of the crown to be sure that subjects lived in peace and were able to carry out their occupation unimpeded by others. When the American colonies became states, they continued to build upon the fundamental rules of the common law. As the economy and society developed, the common law rules were more difficult to apply. What if a pig farm were now a factory, with numerous employees, housed in a concentrated urban area? The annoyance issuing from a factory would now be more widespread. Harm to employees also increased under industrial working conditions, and the old rules proved to be less protective of them as well.[24] In response to modern conditions, the administration of such common law rules was transferred from court rulings to statutes, so that legislatures could identify the needs of the community.[25] At the local level, city councils passed ordinances and set up new departments in the executive branch to carry out the will of the legislature.

That authority was needed as cities became more concentrated and epidemics became more deadly. In the nineteenth century, cities could expect regular outbreaks of disease, and sometimes these could be devastating. In 1878, yellow fever hit Memphis hard, claiming the lives of five thousand people in a city with a population of 33,000, initiating the short-lived experiment of the National Board of Health.[26] At the local level, George Waring, the renowned sanitary engineer, built the first of his Waring sewerage systems, consisting of narrow brick pipes connecting houses to streets. The pipes were dedicated solely to household sewage, which was rapidly directed to a flushing tank and washed away from habitation. Reportedly, it worked. Memphis went "from one of the most unhealthy cities in the United States, to one that 'has now a mortality that will compare favorably with the best in all the land.'"[27] In subsequent decades, the Waring system was put in place in cities large and small all around the country. Waring emerged as a leading sanitary engineer, developing ways to clean city streets and the pipes underneath them.

The response to Memphis's yellow fever outbreak tracks some basic features of governing. A condition can be difficult but tolerable. As living conditions changed, and cities became more concentrated, contagious illnesses spread more effectively. There may have been a response, but whether a governmental response would be federal, state, or local was not readily evident. Which one of those levels of government had the capacity was likewise not clear, and whether it could or wanted to enlist private resources was not guaranteed. Whether that level of government had the political will to actually respond, and respond effectively, was another uncertainty.

The sanitary movement brought a more informed sense of what to do with these institutions. By participating on boards of health, physicians were able to keep them up and running, sustaining them as permanent institutions beyond any health scare. They brought expert knowledge. With the help of sanitary campaigns, these existing boards of health became the site of early municipal garbage collection programs. Disposal of garbage, defined as "refuse, accumulation of animal, fruit or vegetable matter, liquid or otherwise, that attends the preparation, use, cooking, dealing in or storage of meats, fish, fowls, fruits or vegetables," generally fell under the supervision of the board of health.[28] As members of boards of health exercised their authority, however, they ran into a lack of capacity to notify, to inspect, to punish breaches of law, and to actually abate the nuisance. Boards of health lacked the ability to carry out a citywide garbage collection program.

The New Orleans Auxiliary Sanitary Association (ASA) was incorporated to report on the conditions that had led to the epidemic. It found household waste and backyard filth to be "the greatest sources of vitiated air," while the banks of the Mississippi River were "made a common receptacle of, and reeking with garbage and filth of all kinds." The commission condemned as sanitary threats the presence of kitchen offal, backyard filth, and reeking piles of refuse along six miles of the riverbank. The commission recommended that "on stated mornings . . . corporation cars should traverse the streets, for the purpose of gathering up the general kitchen offal, which should be placed by the house-holders near the edge of the banquette in boxes or barrels, to be emptied into the river."[29] Despite the concern, the city did not take measures to prevent another epidemic, leaving New Orleans "the filthiest hole in the land."[30]

Local and state agencies worked with the National Board of Health to combat the threat of another such epidemic.[31] Sanitarians drew attention to dumping grounds described as a "festering, rotten mess" of garbage, containing dead dogs and cats, picked over by ragpickers and pigs alike.[32] New Orleans had regulations to deal with filth and garbage, and passed an ordinance regulating trash collection in 1871, but enforcing them was another matter. Illegal dumping and non-enforcement continued, evident in an 1874 sanitary inspector report and the rebuke of city officials after the 1878 "Yellow Jack" outbreak.[33]

The board of health received complaints about the use of kitchen garbage and dead animals as street fill, enticing hogs to tear the streets apart to recover them.[34] City carts were supposed to haul trash to the garbage wharf. There it was to be loaded onto the barges of a private contractor, towed downriver, and dumped. But garbage collection never covered the entire metropolitan area adequately. Reeking refuse was allowed to stand for days on some streets. Furthermore, the garbage boats broke down several times and lay idle for months. This led to the practice of merely dumping the garbage on some neutral ground in sparsely settled sections or using garbage as fill after canal bank repairs.[35]

But the city still lacked mechanisms to enforce the rules it had in place. Ordinances specified what should be collected, how, and by whom, yet city officials and employees who failed to carry out these regulations faced no punishment. Despite the establishment of boards of health at the time of the wave of garbage ordinances, the board of health in New Orleans was of limited use in actually picking up garbage.

St. Louis had better initial success with government-sanitarian relations. In 1885 the health commissioner reported laudable work under the chief sanitary officer. The health department was given only limited funding, and the department expended it in full during the hot season between April and October. An organization sprang up to fill in the gaps—the Citizens' Sanitary Aid Association, funded by local businessmen. The association quickly raised $15,000.[36] With this public-private collaboration, the city was able to sustain a corps of sanitary officers and inspectors, at least during the hot season.[37] By 1889, the city reported concerns about inadequate dumping, and by 1890, a garbage contract was awarded to an outside company, and the Citizens' Sanitary Aid Association faded from health commissioner reports.

The growing profession of sanitary engineering allowed sanitarians to carry out large-scale public works projects. Sanitary engineers were trained to carry out systematic studies across the country. As the superintendent of health in Providence, Rhode Island, Charles Chapin studied the garbage collection practices of 114 cities in his nearly one-thousand-page volume on municipal sanitation.[38] Recognizing that municipal reports were not uniform across the country, and that navigating them could feel like wandering through a maze, he provided an inventory of existing practices of cities, in all their various sizes, climates, and resources. Trained as a medical doctor, Chapin was conversant with state legislation that established the composition of city boards of health. He was knowledgeable about the biological processes of communicable diseases. And he knew how to analyze the myriad technological innovations that cities had devised to collect garbage (or carry out wastewater or clean the streets). A sanitary engineer had to have the medical expert's concern for public health; had to accumulate the necessary professional experience; connect the legal, legislative, administrative, biological, and technological needs; and be able to conduct comparative studies.

Sanitary engineers could envision large-scale city projects. As street commissioner of New York, Colonel George Waring established a fleet of street cleaners. Administration of street cleaning in New York had been under the purview of the police department until 1881, when the Department of Street Cleaning was established. It operated under contract for fourteen years until it was reorganized, and Waring took over as sanitation commissioner in 1895. He organized the sweeping and removal of refuse on 433 miles of paved streets in the city, instituting novel methods of administration and management of personnel, notably dressed in white duck coats.[39] He also designed citywide sewage systems. Like other sanitary engineers such as Rudolph Hering, Waring's firm continued to serve as a consultant to cities looking to install large-scale sanitation projects (figure 2.1) even after Waring's death from yellow fever, which he contracted while working in Cuba.

Civic Organizations

Civic organizations, too, were available to offer advice or volunteer for municipal projects. In fact, these organizations picked up the slack where government was doing too little. If government was not providing adequate education or welfare benefits, civic organizations would step in to provide them in the social sphere. Civic organizations tended to be organized by the middle class, sometimes with access to well-heeled philanthropists. Both men and women were active members of organizations.

Civic associations proliferated in the late nineteenth century as part of a long American tradition. Alexis de Tocqueville noted the proclivity of Americans to join groups when he visited the United States in

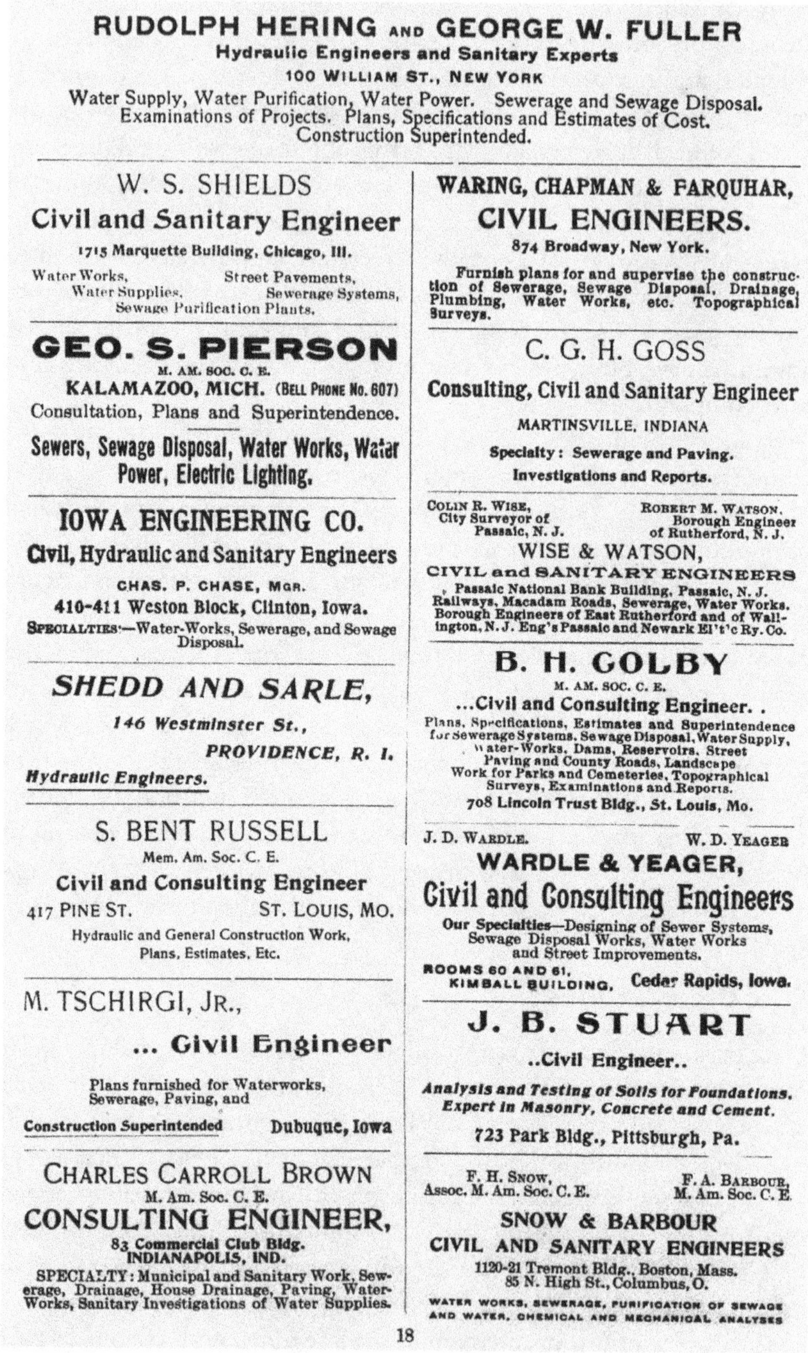

RUDOLPH HERING AND GEORGE W. FULLER
Hydraulic Engineers and Sanitary Experts
100 WILLIAM ST., NEW YORK
Water Supply, Water Purification, Water Power. Sewerage and Sewage Disposal. Examinations of Projects. Plans, Specifications and Estimates of Cost. Construction Superintended.

W. S. SHIELDS
Civil and Sanitary Engineer
1715 Marquette Building, Chicago, Ill.
Water Works, Street Pavements, Water Supplies, Sewerage Systems, Sewage Purification Plants.

GEO. S. PIERSON
M. AM. SOC. C. E.
KALAMAZOO, MICH. (BELL PHONE No. 607)
Consultation, Plans and Superintendence.
Sewers, Sewage Disposal, Water Works, Water Power, Electric Lighting.

IOWA ENGINEERING CO.
Civil, Hydraulic and Sanitary Engineers
CHAS. P. CHASE, MGR.
410-411 Weston Block, Clinton, Iowa.
SPECIALTIES:—Water-Works, Sewerage, and Sewage Disposal.

SHEDD AND SARLE,
146 Westminster St.,
PROVIDENCE, R. I.
Hydraulic Engineers.

S. BENT RUSSELL
Mem. Am. Soc. C. E.
Civil and Consulting Engineer
417 PINE ST. ST. LOUIS, MO.
Hydraulic and General Construction Work, Plans, Estimates, Etc.

M. TSCHIRGI, JR.,
... Civil Engineer
Plans furnished for Waterworks, Sewerage, Paving, and
Construction Superintended Dubuque, Iowa

CHARLES CARROLL BROWN
M. Am. Soc. C. E.
CONSULTING ENGINEER,
83 Commercial Club Bldg.
INDIANAPOLIS, IND.
SPECIALTY: Municipal and Sanitary Work, Sewerage, Drainage, House Drainage, Paving, Water-Works, Sanitary Investigations of Water Supplies.

WARING, CHAPMAN & FARQUHAR,
CIVIL ENGINEERS.
874 Broadway, New York.
Furnish plans for and supervise the construction of Sewerage, Sewage Disposal, Drainage, Plumbing, Water Works, etc. Topographical Surveys.

C. G. H. GOSS
Consulting, Civil and Sanitary Engineer
MARTINSVILLE, INDIANA
Specialty: Sewerage and Paving.
Investigations and Reports.

COLIN R. WISE, City Surveyor of Passaic, N. J. ROBERT M. WATSON, Borough Engineer of Rutherford, N. J.
WISE & WATSON,
CIVIL and SANITARY ENGINEERS
Passaic National Bank Building, Passaic, N. J.
Railways, Macadam Roads, Sewerage, Water Works. Borough Engineers of East Rutherford and of Wallington, N. J. Eng's Passaic and Newark El't'c Ry. Co.

B. H. COLBY
M. AM. SOC. C. E.
...Civil and Consulting Engineer. .
Plans, Specifications, Estimates and Superintendence for Sewerage Systems, Sewage Disposal, Water Supply, Water-Works, Dams, Reservoirs, Street Paving and County Roads, Landscape Work for Parks and Cemeteries, Topographical Surveys, Examinations and Reports.
708 Lincoln Trust Bldg., St. Louis, Mo.

J. D. WARDLE. W. D. YEAGER
WARDLE & YEAGER,
Civil and Consulting Engineers
Our Specialties—Designing of Sewer Systems, Sewage Disposal Works, Water Works and Street Improvements.
ROOMS 60 AND 61, KIMBALL BUILDING, Cedar Rapids, Iowa.

J. B. STUART
..Civil Engineer..
Analysis and Testing of Soils for Foundations. Expert in Masonry, Concrete and Cement.
723 Park Bldg., Pittsburgh, Pa.

F. H. SNOW, Assoc. M. Am. Soc. C. E. F. A. BARBOUR, M. Am. Soc. C. E.
SNOW & BARBOUR
CIVIL AND SANITARY ENGINEERS
1120-21 Tremont Bldg., Boston, Mass.
85 N. High St., Columbus, O.
WATER WORKS, SEWERAGE, PURIFICATION OF SEWAGE AND WATER, CHEMICAL AND MECHANICAL ANALYSES

18

FIGURE 2.1. Sanitary engineers advertise their services
Source: *Municipal Engineering* 23, no. 6 (December 1902): 18.

1831.[40] By the 1890s, Americans both rural and urban formed associations with like-minded people for recreation, for mutual benefit, to volunteer, raise money, or provide services to others. While this work was done in the social sphere, its political import should not be understated. Civic associations are places where ordinary people can learn the skills of leadership and organization that can lead to political participation.[41] In their own right, civic associations have served as an opportunity to participate for those who were disenfranchised. Fraternal associations allowed Black men to gather in support of one another and their community. Similarly, women, both Black and white, joined the club movement to participate and to offer services that government did not.[42]

Some civic organizations rested on shared business interests. In a city such as New Orleans, which was a bustling port that was also susceptible to disease epidemics, sanitarians were keen on issuing quarantines on incoming ships. Quarantines, of course, were bad for business. Businessmen took up sanitation on their own terms. The ASA worked to diagnose the sanitary problems of New Orleans and lobbied for the creation of a board of public works. Members were concerned less with sanitary standards than with clean, passable streets over which commerce could literally travel. The ASA donated three scows to send garbage down the river and rid the streets of loads of trash. The ASA went on to reform New Orleans's drainage system, extend its water supply, and create public bath and washhouses. It offered to take on the funding, implementation, and oversight of garbage collection too. It paid the wages of ten new sanitary inspectors. It bought and covered the city carts as well as garbage boats to take refuse farther out on the Mississippi River to be dumped.[43] The ASA made public appeals for increased attention to the problem of garbage in repeated letters to the editor of the *Daily Picayune*. The ASA was poised to work with the reform mayor, Joseph Shakspeare (who served from 1880 to 1882), and it looked as if sanitation in New Orleans, promoted by both sanitarians and businessmen, would finally make progress. By 1884, however, the "Regulars" were back in elected office. In 1884, the ASA volunteered to take over garbage collection and disposal completely, but the mayor, William J. Behan, refused the offer.[44] Any progress made by sanitarians and the ASA was rendered moot. The board of health reported in 1890 that householders and city employees were dumping garbage into the Mississippi River. Dumping was the preferred method of disposal, as the garbage was expected to wash out to sea, but most of it lingered near the shore, polluting the water supply and unleashing "another revolting odor to join those rising from the sewers and gutters."[45]

Pittsburgh's women's clubs saw active participation by both white women and Black women, operating in separate clubs that rarely combined resources. In the 1890s, Pittsburgh's Black population included a strong middle class with women who were educated and had time to commit. These women's organizations grew up alongside white women's organizations, in keeping with a proliferation of Black women's clubs across the country.[46] In 1894 Rebecca Aldridge invited seven women to talk about the need for a club for self-improvement. They took the name "Belle Phoebe" and sent a delegate to the First National Conference of the Colored Women of America in Boston.[47] Noting the growing number of Black women's clubs, President Josephine St. Pierre Ruffin noted the need for collegiality among "earnest, intelligent, progressive colored women" who could train their children, lift the moral education of Black Americans, and join the causes of temperance, morality, higher education, and sanitary issues.[48]

At this first meeting, the organization issued resolutions, some of which were similar to those proposed by white women's clubs, such as addressing the problem of families living in one room. Clubs could promote mothers' meetings to teach women how to maintain their homes and their privacy. Resolutions also included commitment to both a John Brown Memorial Association and a Frederick Douglass Memorial Association as well as an appeal for an anti-lynching law and a condemnation of the convict lease system. The Pittsburgh and Allegheny chapter singled out the journalist and organizer Ida B. Wells-Barnett for her work.[49]

Members of Pittsburgh's Belle Phoebe League noted that their collaboration allowed them to show their appreciation for one another, and that their interest was solely literary, although they extended their scope. "Our platform is broader," they explained, "seeking no longer to improve ourselves and our own homes, but others."[50] The group's mission was "the mental and moral development of women, and the improvement of the home and race."[51]

The Civic Club of Allegheny County (CCAC) took great interest in the garbage issue, and white women were the drivers. The environmental conditions of the industrial city of Pittsburgh generated disparities between tycoons, the working class, and the middle class. That middle class produced a generation of women who turned their attention to the contaminated air and water and living conditions.[52] A Women's Health Protective Association (WHPA) had organized in 1889 to combat impure water and inadequate garbage disposal. When the city returned to coal as a source of fuel in 1892, the WHPA addressed the smoke problem. The resulting respite from smoky skies gave rise to the demand to clean the city more generally.[53] Across the river, Allegheny had in Mayor William Kennedy a "great enthusiast on sanitary improvements" who thought "the health of the people is the first thing that should be looked after."[54]

When the Republican machine of Christopher Magee and William Flinn captured city government in the 1890s, the reformers teamed up to oppose them. The WHPA assembled a meeting of local preachers, lawyers, doctors, businessmen, and their wives, as well as political and civic leaders, merging men and women into the CCAC. Allegheny's Mayor Kennedy was elected chairman, and he praised the women of Allegheny and Pittsburgh for their charitable and sanitary efforts. A chemist drew attention to the connection between contaminated water supply and city mismanagement but commented hopefully that "the time was near when 'gangs' and 'bosses' would be 'improved' out of existence."[55] In a commitment to clean up the city's streets and its government, the CCAC's first activity was to provide pure water and garbage disposal.[56] The Civic Club claimed that garbage ordinances passed in both Pittsburgh and Allegheny in 1895 owed their origin to the WHPA.[57] The CCAC's effectiveness did not match its enthusiasm, however; it played only a minor role in municipal garbage collection policy. The Pittsburgh machine did not invite it to partner in its garbage collection and reduction programs. Instead, the CCAC stood on the sidelines, carefully watching.

Engineers

Still, there were garbage ordinances, and they had to be carried out. The new need for sanitary provisions and the scale of municipal projects invited another type of expert to participate: civil engineers. Unlike sanitarians, sanitary engineers, and civic organizations, engineers were often invited into city government to carry out plans the cities developed.

Garbage was just one of the public works projects available to engineers at the time. Growing cities required street paving, street cleaning, lighting, gas, electric, and telephone services, sewerage, delivery of clean water, as well as garbage collection. Opportunities for engineers depended on the type of project. Sewers tended to be owned and run as city services, while gas and telephone tended to be contracted out to private companies. Waterworks and electric services struck a fairly even balance between municipal ownership and private ownership. That determination largely rested on the technology. Waterworks that used a rudimentary slow sand filtration system tended to favor city ownership. Those plants that used mechanical filtration were likely to be contracted out. The engineer M. N. Baker explained this pattern as the result of cities being unwilling to purchase the patents for the new technologies.[58]

Whether they were run by the city or contracted out, city government needed to house these services in some agency, existing or new. Treatment of human waste was a long-term concern for sanitarians, as it was a source of communicable disease and general unpleasantness. A backyard privy vault, unconnected to a sewer system, could "render life in a whole neighborhood almost unendurable in the summertime." Privy vaults accounted for three quarters of the complaints registered to boards of health. As cities developed municipal sewer systems and, in those communities without sewers, requirements for privy vaults, sanitarians retained control, and plumbing tended to remain under a city's board of health rather than a buildings department. Waterworks, also a sanitary concern, tended to be run by engineers in a department of public works or special commission.[59]

Driven by public health concerns, sanitarians favored municipal ownership as city governments developed their expertise within city agencies and tightened up their administrative operations.[60] Engineers were more likely to follow a business model, seeing emerging public works as an opportunity to profit off city investment, which was extraordinary. The president of the American Society of Municipal Improvements noted that city expenditures exceeded those of any private industry.[61] Engineers challenged the presumption that city ownership was more effective than private.[62] In garbage collection, sanitarians preferred city-run programs because of the nature of the work. Collectors and drivers had to be familiar with the routes, the proclivities of particular residents, the location of cans, the habits of the community. A permanent force of city employees would develop that familiarity.[63] Engineers were more concerned with the machinery of both collection and disposal.

The engineering profession was changing at this time as well. Whereas an engineer of old may have done all the different types of work on a project, larger projects required a chief engineer to supervise specialized workers and technicians.[64] An engineer might find himself in charge of a water provision project, or sewage, or street cleaning, or garbage collection. These projects would be housed in a department of public works or similar office. Engineers, then, assumed a track alongside sanitarians and learned how to deal with administrative and legal questions.[65]

In 1894, professionals in these positions founded the American Society for Municipal Improvements. The group was comprised of those in charge of municipal departments or public works. The object of the society was to share information about the management of such departments and the construction of public works, to host conventions, to circulate papers and studies, and to disseminate such information in an annual journal. A report on garbage disposal by the director of public works in Fort Wayne, Indiana, pointed to the technical features of garbage collection: cans needed to be watertight and placed in a spot where the collector could reach them efficiently; waste needed to be separated so it could be dealt with properly at the disposal site; the wagon needed to be suited for the method of disposal, and it needed to be accessible to regular cleaning. Such matters required expertise, and such activity had to be sustained. Because of such technological requirements, the report recommended that the collection of garbage should be under the authority of the city, within a public works department.[66]

There was a shared sense among sanitarians, municipal reformers, civic organizations, and engineers that governments could be used to improve the condition of modern cities, and expertise—of technology, of health, and of governing—could sustain clean and adequate governing. Engineers knew which bricks to choose; how to test sewer pipes, paving, and other cement; how to clean streets, collect garbage, and clean and flush sewers economically. Of all the rising experts and reformers reviewed here, engineers gained the most access, and cities did tend to entrust planning and administration to them.[67]

Engineers were first and foremost professionals in their own fields; they were not primarily committed to government reform. The seventh annual meeting of the American Society of Municipal Improvements in 1900 saw only about sixty attendees, and most of the committee reports were the work of just one author.[68] A questionnaire sent to forty-four cities about their garbage collection practices and street cleaning brought only eight responses, whereas the APHA had conducted a similar survey years prior with much more thorough results. The *Engineering News* suggested that reform associations such as the American Public Health Association, the American Society of Municipal Improvements, and the National Municipal League consolidate their efforts and stop replicating their studies and tapping out available resources.[69] Engineers were reluctant, nevertheless, to defer to the conclusions of sanitarians. The former suspected the latter of operating on personal opinion or of having commercial interests in one method or another. Engineers feared that "the whole garbage disposal problem has been and is largely still being approached altogether too much from the viewpoint of odors and sentiment and too little from that of modern sanitary science."[70] City engineers networked with other reformers, but they maintained their own professional boundaries and claimed their territory in public works. Cities hired civil engineers directly, and they contracted out engineering projects, so engineers had access that other reformers lacked. For them, municipal problems arose when there was not enough engineering; they were not concerned about corruption or the structure of government. If they were bothered by municipal mismanagement, it was because of the failure of the public and non-technical city officials to realize that "the most difficult of the problems involved are engineering in character and will never be satisfactorily solved until they are entrusted to engineers."[71] They were interested in developing their own technologies and using municipal innovation to land large municipal projects.

The journal *Paving and Municipal Engineering*, devoted "specially to street and sidewalk paving, sewerage, water-works construction, street lighting, and all matters pertaining to municipal engineering," began publishing in 1890 after organizers of an exhibition of street paving from across the country found that there was no medium for communicating with all pavers. Apparently, pavers had been finding one another through advertisements. The convention gave rise to the publication as a way for those in the paving business to network, which broadened to include those in public works more generally.[72] Engineers did not just want to secure projects; they wanted the work to be done well, for the improvements to be of "permanent and durable character, obviating the necessity of frequent reconstruction."[73]

Garbage collection and disposal were standard improvements for engineers, especially the more ambitious projects of reduction. An engineer could follow along with studies and existing projects by reading the many articles published in *Municipal Engineering*. An engineer could keep track of cities that were considering reduction or other disposal facilities through notices in the journal that reported when cities were inquiring about facilities or when the board of health voted on, or even considered, an incinerator.[74] The engineer who kept up with the profession would know when a city was calling for bids on contracts or if a city needed to hire an engineer.

This networking allowed engineers to exchange information about even the most obscure features of garbage collection. Garbage cans were prime sites for innovative adaption for citywide collection programs. One inventor developed garbage cans with tapered sides so that they could nest inside one another for storage. A woman inventor topped that with tapered sides and interior stops so that cans could be nested without wedging.[75] Garbage wagons, which traveled through streets and likely smelled (if they did not have garbage spilling out of them as they moved to the horses' gait on uneven streets), deserved attention, and engineers thought about them considerably and carefully. Articles and advertisements served to announce innovations in garbage wagons and other collection apparatus.[76]

The material for carts, whether they were covered, how they had to be designed to collect garbage most efficiently, their design for the point of disposal, how easily and often they could be cleaned—determining all of this took the expertise of an engineer (figure 2.2). Or a garbage cart could be designed to bring the entire can to the disposal plant, eliminating the need for messy dumping on the sidewalk.[77]

FIGURE 2.2. Sanitary garbage cart
Source: *Municipal Engineering* 52, no. 6 (June 1917): 306.

The Garbaget was an idea for a vehicle that would travel around the city (figure 2.3). Garbage could be fed into it and turned into a dry powder.[78] The entire apparatus consisted of a cylinder on the outside with swinging hammers that struck the teeth of an inner cylinder, which would pound the garbage. Two smaller shafts with hammers would mix the mass. A perforated plate with one movable end separated the compartments. Divided garbage would pass through the perforations in the plate. Gas could escape and liquids be drained out. The garbage would be pulverized so that it would be odorless. The Garbaget could operate as it traveled down the street. The apparatus never took hold, but the opportunities for municipal funding encouraged such innovative thinking.

Once garbage was carted away, engineers had more opportunities in designing disposal plants. Sanitarians had their favored methods, namely incineration and reduction, but they conceded that the method of disposal depended on the city—its climate, its resources, and its

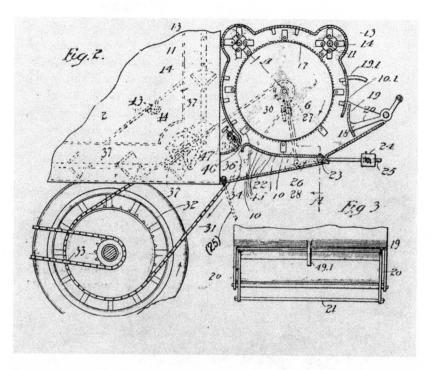

FIGURE 2.3. The Garbaget
Source: *Municipal Engineering* 52, no. 1 (January 1917): 34.

political will. Because of the variety of methods, engineers had a variety of projects to participate in. Their expertise would be needed in deciding how to sort household waste, to determine if fills needed to be covered, and if so, with what material.[79] They also could get involved with innovative developments, such as reduction—extracting steam power or fuel from the processing of garbage.

Reduction was an elaborate process in which raw garbage was placed in an airtight tank. The tank had walls and a jacketed bottom to control the temperature and allow steam to be produced without coming into contact with the garbage. The top of the tank closed, and the heat from the steam reacted with a solvent in the garbage, vaporizing it. The solvent and water were then separated, with the solvent returned to the tanks and used again in the reducers. When the garbage was thoroughly dried, the solvent, containing grease from the garbage, was drawn off into an evaporator, vaporized again, and transferred to storage tanks. The remaining degreased waste could be used for fertilizer.[80]

Engineers' specialized knowledge made them less expendable than some of the other reformers and professionals of the time. If they were ignored by their local government, they could muster their collective power as a profession.[81] Engineers reinforced their resources by holding their own professional conventions, allowing for the dissemination of knowledge and insider information. If cities did not hire them directly into their departments of public works or departments of streets, engineers were poised to contract as private parties with governments.

As important as engineers were to running garbage plants, they were not always consulted about which garbage program to choose, because science was not used to make political decisions. Recognizing that the problems of garbage disposal were "largely technical," the sanitary engineers Rudolph Hering and Samuel Greeley hoped for "systematizing the entire work under competent engineering advice and management, with the intimate co-operation of the health officers, and, under carefully drawn ordinances, regulating such community work."[82] Cities had streets, but the street plans were poorly executed. The streets may not have been graded properly, or cities may have chosen poor paving materials, or may not have kept up with maintenance. The same went for waterworks, milk distribution, lighting, and garbage. As the associate editor of *Engineering News* put it, garbage collection was "strangely neglected in the majority of cities and towns, and its final disposal is one of the greatest blots upon American municipal administration." The problem, for engineers, was that cities were making technical decisions through a city council or committee composed of laymen, who would tour the facilities of another city, meet with the commercial agent of the plant, who would wine and dine them, and, on the basis of that visit, they would make their decision. Cities were not making decisions on a "scientific basis." While larger cities had made some attempts to use technology to collect and dispose of garbage, they were not always successful, and smaller cities were doing little if anything.[83]

City officials might not realize what conditions would make a project appealing to an engineer. If a city were to contract out the creation of a facility, officials might not know how long the contract should be for the engineer to recoup his investment. The average length of contracts in 1902 was three years, not long enough for the contractor to see a return on investment. Furthermore, maintaining a facility requires experts. Furnaces needed to be run at the proper temperature. Pumps and filters needed to be maintained.[84]

Municipal Reformers

Progressive Era municipal reformers used the tools of social science to question the structure of city government, its goals, and its operation. The concentrated populations, cramped living quarters, poor working conditions, and poverty of urbanization prompted responses including scientific study and formulation of policy. In addition, reformers worked on improving city government itself, challenging corruption, instituting civil service and administrative efficiency, and populating its offices with experts from those social science fields. To combat the concentrated power and influence of political machines, municipal reformers looked to structural changes in city government. Changes might include a strong mayor (to decrease the influence of the machine on city council members), professionals placed in agencies, merit-based appointments, home rule charters, and city managers.

In 1891 the first issue of *Annals of the American Academy of Political and Social Science* presented studies of and recommendations about municipal government. A supplement to its first issue included an article titled "Public Health and Municipal Government." Pointing to the rapid increase in population in cities, it singled out newly arrived immigrants, reflecting prejudice against ethnic immigrant groups: "These people congregate in certain quarters and in certain houses which are adapted to their means, tastes, and habits—they huddle together in foul rooms; they include the loafers, the street arabs, the tramps, and casuals; their poverty is the result of intemperance and indolence dependent on physical structure." Nevertheless, the article noted paternalistically, "we must look after these people, and help them, for the sake of others, if not on their own account."[85] In responding to the challenges of urbanization, reformers relied on classifying residents by status, opening the opportunity for future public health campaigns targeting poor neighborhoods to deliver their services, including garbage collection.

Municipal reformers multiplied their resources by networking with one another. In 1894 the first National Conference for Good City Government met in Philadelphia for the purpose of "stimulating and increasing the rapidly growing demand for honest and intelligent government in American cities, and to discuss the best methods for combining and organizing the friends of Reform so that their united strength may be made effective."[86] That conference gave rise to the National Municipal League, formed later that year. Its purpose was "to multiply the numbers, harmonize the methods and combine the forces of all who realize that it is only by united action and organization that good citizens can secure the adoption of good laws and the selection of men of trained ability and proved integrity for all municipal positions or prevent the success of incompetent or corrupt candidates for public office."[87] Through annual meetings and publications, the National Municipal League relied on local members to investigate corruption and poor governing in their own cities and share that information with the league. It could both expose mismanagement and recommend better practices.[88]

Municipal reformers sought to shape the organization of city government to remove corruption and design administration to carry out effective delivery of services. Sanitarians joined them, weighing in to offer their expertise on how best to design public health agencies. Dr. John Billings spoke at the National Conference for Good City Government, advising that if reformers wanted to staff boards of health, they needed to hire men with years of training in medical, statistical, bacteriological, chemical, and legal knowledge, and with the ability to discuss public works plans. They should also be given sufficient authority to do their jobs. Board of health officials could run into resistance from engineers or police officers they had to work with. Billings recommended that the board of health report directly to the mayor.[89] Yet nearly a decade and a half later, George Soper's study of local boards of health found that they did not uniformly have the methods to carry out their basic functions, such as collecting vital statistics. "To be effective health work must be cooperative," with experts able to collect the information they need and work with officials across levels of government.[90] Furthermore, the staffing of boards of health should be removed from political patronage and filled with trained professionals.

In discussing Soper's paper at the National Conference for Good City Government in 1908, Dr. J. F. Edwards, of the Pittsburgh board of health (under Mayor George Guthrie, who was a founding member of the National Municipal League) picked up on Soper's attention to the need for public education, whereby individuals could practice hygiene and sanitary practices that could prevent sanitary problems. Edwards noted the enlisting of "a vast constituency to supplement and aid official work . . . bringing them into closer touch with the public."[91]

Municipal reformers recognized that their efforts needed to be internalized by the public. Public adoption required more than information gathering and expert advice; it required interaction with the public. Coordinating with sanitarians, reformers offered to design governments that could make use of sanitarians' expertise, only to run into obstacles posed by patronage and lack of will among political regimes.

Reformers Rebuffed

Despite resounding support for municipal collection programs and the attendant development of expertise, reformers and experts found themselves left behind by municipal governments as they carried out new garbage ordinances. Engineers were well situated to lend advice, but even they expressed frustration that cities were going about making decisions on garbage collection and other municipal programs not based on science. The other experts—sanitarians, municipal reformers, and civic organizations—were sidelined to various degrees, even as garbage collection programs were adopted. These organizations networked with one another, sharing information and advancing opportunities.

Decisions about sanitation were made not by sanitary experts but by political officials seeking to deepen their advantage. Corrupt cities found plenty of opportunities to collect garbage without relying too much on these experts. Expert organizations did not give up. They kept track of developments in sanitation and closely watched its progress or shortcomings. They studied it. They exchanged ideas. When corrupt regimes faltered, they would be ready to jump in.

CHAPTER 3

Ready to Profit

Inadequate Garbage Collection by Corrupt Regimes

In the late nineteenth century, cities across the country faced a growing "garbage problem" as trash piled up in vacant lots and streets. Sanitarians, sanitary engineers, civic associations, civil engineers, and municipal reformers were ready to initiate and shape municipal garbage collection programs. They offered help and advice. Yet cities did not rely on the expertise and reformist ideas as the experts and reformers would have liked, and many cities that did address garbage did not choose to do so merely because it was a serious problem. If garbage piling up and expert attention did not lead officials to address sanitatation, what did? Corruption—not genuine concern for public health—created political will to take up the garbage problem when municipal actors found a way they could benefit from it.

Late-nineteenth-century cities had formal democratic institutions—elected officials, city agencies. At the same time, many also had informal regimes that cycled into and out of power. Although corrupt regimes all abused power for personal or party gain, they differed in how corruption was organized, the relationship between informal (corrupt) regime and formal democratic institutions, and the mechanisms by which they exercised it.[1] St. Louis was controlled by a business oligarchy organized outside of government, exercising influence through boodle (bribery of city officials). New Orleans was run by a weak political "Ring" cycling in and out of government, using patronage (putting favored constituencies on the city's payroll) to hold on to tenuous political support. Pittsburgh was ruled by a political machine firmly entrenched within city government, and it used rigged contracts to enrich itself. Charleston had an informal aristocracy in place for decades, and it relied on nepotism to direct city dollars to elites.

Although corruption in all four cities looked different, it similarly motivated elected officials to address the garbage problem when they found it could be politically expedient. City officials chose the method of collection and disposal that would benefit sitting regimes politically or financially. As expected, however, corruption did not always lead to good results. In fact, in many cases it hindered good policy programs. In this chapter we highlight how corruption created the political will to take on the garbage problem while at the same time it led to failure in the creation of durable sanitation infrastructure.

As in other cities across the nation, trash was a problem in St. Louis and New Orleans. In the early 1890s both cities contracted out for garbage collection and disposal through state-of-the-art reduction facilities. St. Louis successfully built a reduction facility but subsequently lost the capacity to use it, while New Orleans's plant never got up and running. Corruption can motivate political officials to address public problems even as it can be a roadblock to good policy results. Accordingly, corruption was an obstacle that thwarted lasting solutions (St. Louis) or prevented innovative solutions from being put in place at all (New Orleans).

Corruption in St. Louis and New Orleans was located outside of government; corrupt regimes were not integrated into formal democratic structures. In St. Louis, power resided purposively outside government in the hands of businessmen, called the Cinch. They relied in turn on the unscrupulous middleman Ed Butler, an Irish immigrant and blacksmith, to bribe government officials. When a new mayor tried to rid the city of corruption (i.e., Butler), he also rid it of the essential capacity to collect and dispose of trash. Power in New Orleans, by contrast, was diffuse. Leaders of New Orleans's Ring wanted to integrate their informal corrupt regime with formal democratic institutions but they were not strong enough to carry out the scheme. Instead, the Ring relied on patronage in city services, including trash collection, to build support in a fractured political system. The Ring had little capacity to design or deliver effective trash collection and disposal, and a state-of-the-art disposal plant never got off the ground. Both cities struggled for decades without adequate garbage services, and into the twentieth century, they both dumped trash on land and in the Mississippi River, a mode of disposal considered "primitive" even at the time.

The Nonintegrated Regimes of St. Louis and New Orleans

Unlike in Pittsburgh and Charleston (chapter 4) where government structures and informal regimes were integrated so that there was broad overlap between the regime's priorities and capacity and the city's, in St. Louis and New Orleans, formal democratic structures and informal regimes were not intertwined. These nonintegrated regimes had a more difficult time building political will and maintaining ability to collect and dispose of waste.

City government in both St. Louis and New Orleans was organized into wards, which in theory provided broad representation, but in practice allowed many places for corruption to fester. In 1875 Missouri adopted a new constitution, and St. Louis—then the nation's fourth-largest city—took advantage of the opportunity. The city separated from the county, and it drafted a charter.[2] The new city government had a legislature with two houses, one elected at large and one elected with members from each of the city's wards. Whoever influenced the city's many elections could also influence the city's policies.

Similarly, New Orleans was made up of many wards. And, as in St. Louis, corruption infiltrated democratic politics. The Civil War had destabilized ruling coalitions, and the city came to be governed by competing factions of the Democratic Party. Elections seesawed back and forth between business-elite reformers and working-class (notoriously corrupt) "Regulars." Neither the reformers nor the corrupt political officials known as the Regulars were committed to equality, but both courted Black voters. Even when they won election, it was hard for either reformers or Regulars to maintain control over the city's decentralized and influential ward leaders. Aspiring political leaders tried to stabilize politics and gain long-term support. The reformers wanted to accomplish these goals through clean government, but the Regulars' strategy was to distribute patronage through public works programs, especially garbage collection.

In both St. Louis and New Orleans, corruption was the motivation for the city to address sanitation. In both places, corruption was located outside of formal democratic institutions. In St. Louis, power lay in the hands of Cinch businessmen, who bribed city officials to get what they wanted, while in New Orleans it was in the hands of the Regulars, who cycled in and out of government. Although the actors and events in each city are different, the pattern is the same. Garbage was a growing concern for both cities. Yet corrupt governments embraced garbage collection and disposal programs only when they saw the potential for political or financial gain.

St. Louis Boodle

St. Louis had formal democratic institutions at the same time it was informally ruled by an oligarchy of businessmen. The city's corrupt businessmen used boodle, or bribery, to persuade political officials to give them favorable contracts. They relied on a middleman, Ed Butler, who exercised a great deal of influence in his political ward—much of it illegal. Butler engineered electoral results using "repeaters," dead voters, pre-printed ballots, and corrupt election judges. In 1876, when urban residents voted to "divorce" the city of St. Louis from the county, they created an independent St. Louis freed from excessive state influence as well as duplication in administrative jobs between the city and the county.[3] Butler found new ways to profit. The local newspaper described the opportunities the new legislature provided Butler:

> Twenty-eight men, expected to furnish capacity and patriotism at $25 a month! How well his plan would work in a city whose growth was opening privilege and profit in every form should have been, it would seem now, not difficult to see. One might have expected that a leader would have sprang [sic] up from within the ranks of the delegates themselves. But the man who never desired to hold office himself saw a way he liked better. "What should I want to hold the office for?" he said in later life. "I always preferred to let the other fellow hold office, and then get acquainted with him.[4]

Butler created "combines" in the assembly, which "received and distributed bribes in exchange for the passage of legislation desired by businessmen."[5] Members of the assembly came to see the use of bribes as valid. There was a price for everything in St. Louis, even legitimate city services for legitimate citizens, such as a street opening, could bring legislators $50 or $100.[6]

Butler essentially created a shadow form of governing that worked alongside and through official political channels. It subverted democratic governing and shifted power from democratic institutions (the assembly) to undemocratic ones (Butler and his combines). Although some people considered Butler a political boss, the organization that he created and maintained did not look like a traditional political machine. Butler did not occupy a formal political position in government. He did not work with only one party. In fact, his apparatus was for sale to the highest bidder.

Corruption in St. Louis was extractive. The "winners" in St. Louis were the Cinch, the business elite who bought the rights to provide lucrative city services (railroads, trolley lines, lighting companies); the corrupt political officials who made ten times their annual salary (or more) in bribes; and Ed Butler, who exercised great power and extracted great wealth. The "losers" were the residents of St. Louis. The exchange of bribes for city services was not limited to wealthy businessmen and big services; even the most minor of city functions, such as a street opening, were fair game for bribery.[7]

New Orleans Patronage

New Orleans too had formal democratic government. Yet it was ruled by a corrupt and relatively weak political ring of Regulars. In New Orleans, patronage was dispersed liberally by Democratic officials to build support in the city's many wards. The department of public works—which spent thousands of dollars a month—was a key agency for rewarding allies. Electioneering was part of the strategy. As a result, elections like that of 1884 were described as "a mockery. The popular will was nullified, the popular voice was stifled and free citizens were robbed of their dearest rights. The people cast the ballots, but the [Regulars] counted them."[8] Electioneering, however, was not foolproof. If it helped the Regulars win office, it could hardly help keep them there. The heart of the Regulars' strategy was patronage through municipal departments like public works.

In 1880 John Fitzpatrick, the "Big Boss of the Third Ward," became the administrator of improvements and, two years later under the new charter, was elected commissioner of public works.[9] A typical page from the payroll of the department of public works lists names of employees in elegant cursive handwriting. Yet the records are striking for how they are organized: by wards where employees worked and resided rather

than by last name or occupation. For example, on one page of the public works payroll for June 1883, the department grouped employees
under the headings "4th and 5th Ward Bridge Gang," "4th Ward Carts"
(which included garbage carts), "5th Ward Street Gang," and "5th Ward
Carts." These headings indicate what type of work is being done (e.g.,
work on bridges by the "Bridge Gang" or work on streets by the "Street
Gang") and where the employees who engage in it live (Fourth and
Fifth Wards). Census records show a near perfect match between where
carts were assigned and where the people assigned to those carts lived.
This single page is similar to the pages that come before it, which cover
the First through Fourth Wards, and the pages that come after it, up
to the Sixteenth and Seventeenth Wards. In fact, this particular page
in this particular volume looks remarkably similar to every other payroll from the department of public works. Although administratively it
may make more sense to organize alphabetically, keeping these records
in this particular fashion allowed Fitzpatrick and other city officials
to show ward leaders just how much they were doing for their favored
constituents.

It turns out they were doing a lot. The department of public works
spent more than $15,000 in the month of June alone. Garbage and
other kinds of carts were paid $2 per day, and the department hired
eighty carts to work an average of twenty-two days. In total, the department of public works spent $7,600 on street "labor" (the quotes are in
the original) and $3,400 on carts.[10] The money spent and the people
employed, however, were designed to build support more than clean
up the city.

Corruption in New Orleans was redistributive. It took money from
city coffers and gave it to the Ring's ethnic working-class constituency,
including widows. While the redistribution may have helped some
of the city's working-class residents, it did not promote better, more
efficient garbage collection and disposal. In fact, it hindered sanitary
advances.

Garbage as Public Health Concern

As in other nineteenth-century cities, the "garbage problem" in both
New Orleans and St. Louis was real and, more than a public nuisance, it
was deadly. Health officials and sanitarians were rightfully concerned.
Yet it wasn't concern for the public or the public's health that motivated city officials to take up the garbage problem. Instead, St. Louis's

Butler saw sanitation as an opportunity to make money, while New Orleans's Ring used it as way to build political support.

Cholera in St. Louis

Public health crises became a means for St. Louis's Cinch to enrich itself, especially Ed Butler. When a cholera epidemic struck St. Louis in 1866, killing more than 3,500 people, public health was pushed to the forefront of the government's agenda. The city created a board of health with the power to draw up sanitary regulations.[11] By 1871 the health department was spending nearly a third of its budget on the removal of offal (animal entrails) from the city's sewers and the use of scavenger boats that dumped the city's waste in the Mississippi River.[12] Various municipal departments were involved with the administration of garbage: the police, the board of public health, the Board of Public Improvement, and the Scavenger Department. But these efforts were not enough. Dumping garbage was unsanitary, and residents complained about the "sickening smell" pervading the entire neighborhood where the scavenger dump was located.[13]

Surely sanitarians, who studied and meticulously chronicled best practices in sanitation, had suggestions about what to do. As chapter 2 shows, their suggestions would likely have improved sanitary conditions in the city and in the lives of its residents. But expert knowledge and advice did not drive policy in St. Louis. As important as proper sanitation was for good health, city officials were more concerned about political calculations, and Ed Butler was the key.

As the conduit between the St. Louis Cinch and political officials, Butler was willing to sell a lot in St. Louis. He took a special personal interest in garbage. Butler used his position to reward family businesses and to line his pockets. After the new charter was enacted, he traded favors with the new mayor, Henry Overstolz. In return for Butler's electoral support, the board of health awarded Butler's brother-in-law James Hardy the garbage collection contract even though Hardy was the highest bidder by $17,000 and the bid he submitted to the city was nearly triple the actual cost of collecting trash.[14]

Even with laws on the books and paid collectors, city streets were not clean. City officials recognized the health implications associated with unsanitary conditions. According to Chief Sanitary Officer Charles Francis, "It is now well understood, and a mass of statistics from health authorities prove it, that many of our most common diseases can be, if not almost entirely prevented, very much restricted." But the barriers were high: in addition to laws and collectors, St. Louis needed compliance from its residents and active programs. Leaving waste in yards, vacant lots, and streets was clearly illegal, but the chief means of addressing the problem was passive, even convoluted. The health commissioner described the process as follows:

> The sanitary officer notifies the Health Commissioner; the Health Commissioner notifies the Board of Health; the Board of Health by order directs the party responsible to appear before it on a certain date, and this order is required by law to be served by the City Marshal in the same manner as writs and summons are required to be served in civil cases; then the Board hears the statement of the case and the matter is condemned as a nuisance; the Health Commissioner then issues an order on the party to abate the same, and the law allows five days in which the party may appear before the Board to protest against the order of the Health Commissioner, so from the first time the notice is given to the Health Commissioner of the existence of a nuisance until the expiration of the time given in the order to abate it, fully four weeks has elapsed.[15]

Furthermore, when trash was actually picked up, the city disposed of it by dumping, not a very remarkable or advanced solution. Health and sanitation officials pushed for better answers, describing the current mode of collecting garbage and "conveying it to a few dumps and there discharging it" as "very objectionable." All manner of waste was hauled through city streets, no doubt giving off noxious odors and spilling from the vessels that should have contained it. Human and vegetable waste was dumped in the river, where it would stay near the shore and "lodge in greater or lesser quantities." Dead animals were taken across the river to a rendering establishment. City officials worried about how long they would be able to keep this up.[16]

There ought to have been a better way to solve the problem. Sanitarians across the United States were concerned with best practices of trash collection and disposal to keep cities clean and disease at bay. Health officials in St. Louis, too, asked the big questions about what to do. Still, St. Louis officials did not know what the best practices might be without making their own inspection of "the processes in operation."[17]

While health and sanitation officials struggled with finding the best—most efficient, most sanitary, most advanced—solution, corruption provided another logic, which pointed to a different set of answers.

The best responses to the city's sanitary problems were the ones that enriched its powerbrokers, the Cinch and Ed Butler. Indeed, garbage collection and disposal became a financial windfall for Butler under Mayor Edward Noonan (1889–1893), a political ally of Butler's, and the following two administrations under Mayors Cyrus P. Walbridge and Henry Ziegenhein.

Yellow Fever in New Orleans

Although the depth of corruption in St. Louis is staggering, of all the cities in this book, New Orleans may have suffered the least sanitary conditions. The Crescent City had few paved streets, so most roads were impassable for garbage carts during bad weather; the summer sun was unforgiving, beating down on New Orleans's uncollected trash; and when waste was removed, the city's policy was to dump it, first on land and later in the Mississippi River.[18] Yet the corrupt New Orleans Ring did not respond to the growing garbage problem and disease outbreaks; instead it used garbage—and public works more broadly—to shore up its rather weak political position.

Perhaps unsurprisingly, more than any other American city, New Orleans suffered from yellow fever, which threatened to spread across the region as ships passed from New Orleans up the Mississippi River.[19] Sanitarians claimed that the city's poor practices contributed to the spread of disease. Indeed, New Orleans used offal, dead dogs, and waste mixed with ashes as fill to raise the grade of streets, "rendering them passable." But rotting garbage and animals were not adequate materials to undergird streets in "a great semi-tropical city."[20]

In response to a yellow fever epidemic in 1878 (which New Orleans concealed), Congress created the largely toothless National Board of Health to help state and local governments clean up their cities and eradicate deadly contagious diseases.[21] Crescent City business leaders organized the New Orleans Auxiliary Sanitary Association, which provided the city with garbage carts and boats so it could now collect and dump trash in the Mississippi River, an improvement over street fill.[22] Both the national board and the sanitary association quarreled with the state board of health, but even if they hadn't, the voices of health officials would still have been drowned out by the single-minded focus on garbage not as a sanitary issue so much as a political one.

New Orleans's garbage carts were cast as a form of charity, given to widows who had few other options for earning a living. According to the

New Orleans Times-Democrat: "Under the plea of charity the custom has been to employ a number of carts belonging, or said to belong, to widows. . . . [T]hese widows were in more or less distressed circumstances, and . . . aid thus given them would go a long way toward their support." Testimony by a foreman working under Fitzpatrick, however, showed that "the widow's cart system which has been carried on at the expense of the city under the pretense that it was a charity, is largely a fraud, a mere trick to allow the ward favorites and political strikers to draw pay without attracting too much attention. Some of the alleged widows turn out to be expert workers at the primaries and ballot-boxes."[23]

The system of garbage collection in New Orleans was truly striking. Although page after page of the payroll of the department of public works lists the names of women as the cart owners, there is little to no evidence that they actually picked up trash.[24] Cart owners were always white. Male owners tended to be married and of German descent. Women tended to be widowed and of Irish descent. Rather than being a form of charity, the system of "widows' carts" allowed city officials to put particularly sympathetic and/or well-connected women on the municipal rolls and take others off. Newspaper managers, district criminal court judges, and superintendents of districts wrote letters asking that carts be added to or removed from the city rolls. A note in the register of city carts for 1900 reads, "Reinstate Mrs Walsh 2326 N. Rampart Street in place of Mrs Krauz," and over it is written, "Order issued May 11, 1900."[25] Sometimes women would write on their own behalf. Mrs. Olivia Babad of the Third Ward, who had recently been added to the Fly Gang, asked, "Could you possibly give my cart as much work as possible, I would not ask so great a favor, only I am depending for a living on same, & if I make only half time, it will only pay expenses."[26]

Building Capacity

While city agencies may have had the capacity to collect trash, they did not have the capacity to dispose of it successfully. St. Louis was able to contract out to Cinch businessmen to build state-of-the art facilities to dispose of trash, but New Orleans was not able to do even that.

Corrupt Contracts in St. Louis

St. Louis's political conduit, Ed Butler, took a personal interest in garbage collection as a way to enrich himself. Although at first Butler (through his brother-in-law) had a lock only on collection, in 1890 the Cinch, Butler, and their political allies sought the same for disposal, too. The city authorized the mayor and health commissioner of St. Louis to contract for the sanitary disposal of slop, offal, garbage, and animal matter in the city by the Merz reduction process (figure 3.1), to be "located at suitable and convenient points within the city limits." To ensure that the contractor actually burned the trash, he was prohibited from dumping in vacant lots, sinkholes, or the Mississippi River, or feeding the waste to hogs. The first garbage disposal contract, for at least ten years, was awarded to the St. Louis Reduction Company, whose president was the Cinch's John Scullin.[27] The St. Louis Reduction Company had a cozy relationship with Butler. He was one of the contract guarantors, a stockholder of the Reduction Company, and on its payroll.

In 1893 the city contracted directly with Butler (and not through his brother-in-law) for garbage collection, "to remove all garbage, dead dogs, cats, rats and fowls from the streets, alleys, roads, courts, private places, private streets, courtyards, market houses, hotels, restaurants, tenement houses, private dwelling houses and public institutions within the limits of the city of St. Louis" twice a week in the fall and winter and three times a week in the warmer spring and summer months.[28] Not only was there more waste in the summer, but also waste left in the hot summer sun caused more problems, which was common in southern cities.

In addition, Butler was required to remove "all garbage which may be offered to him, or to which his attention may be called," whether in alleys, on sidewalks, or behind unlocked doors or gates, but he was not required to remove garbage mixed with ashes, cans, glass, or straw. Butler was to deliver all of this garbage to the St. Louis Reduction Company. He was awarded a five-year, $425,000 contract. The president of the St. Louis Reduction Company, John Scullin, as well as J. B. Clement, president of the St. Louis Sanitary Company (the new reduction assignee), appear as Butler's guarantors.[29]

Contracting out did not solve the city's problems. Trash still piled up in alleys and streets. It continued to be dumped in the Mississippi River. Butler and the Sanitary Company were not doing their jobs adequately. Nevertheless, citizens, city officials, and Butler himself disagreed on who was to blame. Citizens complained about Butler: he failed to collect garbage that was offered to him; he failed to patrol areas; and he dumped garbage into the Mississippi River. They complained about the Sanitary Company: it received credit for disposing of

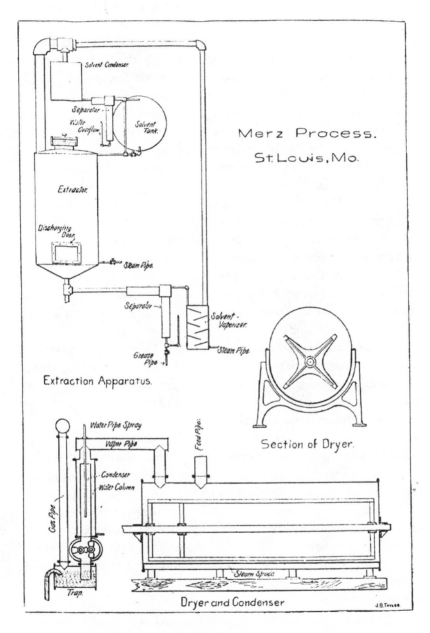

FIGURE 3.1. Merz reduction process in St. Louis
Source: Charles V. Chapin, *Municipal Sanitation* (Providence: Snow & Farnham, 1901), 705.

garbage through reduction when it actually dumped it in the river; and it refused to accept garbage, which was then just dumped in the river too.[30] The Sanitary Company was unable to dispose of all of the waste that Butler delivered to it, and residents living in the vicinity complained of "obnoxious" smells emanating from the reduction works.[31]

City officials defended, and even praised, Butler and the Sanitary Company. The health commissioner commended the company for the condition of the city: "Never within the history of the City, as far as remembered, has there been such a perfect condition of sanitary and healthful cleanliness. . . . The Health Department does not hesitate to attribute this desirable condition of affairs directly to the work of the St. Louis Sanitary Co." The officials focused on the improvement by comparing Butler and the Sanitary Company's performance to the alternative: "The enormous amount of garbage, butcher's offal and other animal and vegetable refuse disposed of and destroyed by this company would be appalling were said amount diverted as formerly into the river or allowed to remain half collected and half hidden in the alleys and byways, as was the case before this company commenced its operations."[32] Certainly Butler and the Sanitary Company were better than nothing. But were they doing their jobs adequately? And was this a good use of city funds?

Both Ed Butler and the St. Louis Sanitary Company acknowledged that they might not be picking up all of the city's waste. They shifted the blame to the city, however, noting they were being asked to do things that were not in their contracts. When Butler removed waste from slaughterhouses daily although his contract required him to do so only three days a week, technically he was doing work that he had not contracted to do. The city counselor informed the board that under the current contract, the city would have to pay for the additional days. Likewise, the St. Louis Sanitary Company was destroying an average of 165.4 tons of garbage per day, sixty-five tons more than the one hundred tons per day for which it received compensation. City officials praised the company, noting it "has at no time evaded its duties and obligations, and has even on many occasions waived its technical rights in deference to this department, where its contract did not, in the letter, oblige it to take action."[33]

City officials in turn put the blame on citizens. The health commissioner noted in a report in 1901 that garbage is "generally well collected," but when there are ten thousand with so many families throwing garbage in streets, alleys, and lots, there will be accumulation, some of which Butler collected and some of which he did not. The city collected 340 tons of garbage daily, 270 tons from homes. Yet less than 5 percent of household garbage was put in the proper place, at the proper time, in the proper container: "Four and one-fifth percent of 270 tons is about 11 tons, which is the amount which the contractor would be required to take if the terms of the contract were strictly followed out, and the remaining 259 tons would be left in the alleys or on the streets."[34] By contrast, the city collected seventy tons from businesses, such as hotels, restaurants, and slaughterhouses, and that waste was generally placed where it ought to be, when it ought to be, and how it ought to be.

The position of city officials, who were well connected to Butler, was that garbage collection might not be as thorough and complete as it could be, but "it is exceedingly well collected when it is considered how few of our citizens comply with the requirements of the City in relation to the placing of their garbage." And while the board of health had the power to make Butler carry out his duties, it could do little to compel the city's residents to do what was expected of them: "Until the citizens of their own volition, or until they are compelled by a very strict law to properly place and care for their garbage, the work of collecting the garbage will not be performed satisfactorily, and it will be impossible to strictly enforce the provisions of the contract." The commissioner's report concludes, "Considering all the circumstances, the removal of garbage by the contractor has been well done."[35]

Patronage in New Orleans

While St. Louis was building capacity for garbage disposal, New Orleans never got that far. The Ring's tenuous political position made it more difficult to change existing garbage collection practices. New Orleans voters swept the corrupt Regulars from office in the election of 1888 and replaced them with reform mayor Joseph Shakspeare, who had served as mayor once before (1880–1882). Shakspeare wanted to address trash in a new way, and he set out to discover more about the garbage problem by traveling to other cities—New York, Chicago, Memphis, Nashville, Louisville, and Cincinnati—to learn what they did.[36] The city considered an ordinance to collect trash and dispose of it by cremation, but health officials worried about the stench involved.[37] For all of Shakspeare's reform efforts and his desire to find ways to make the city better, he struggled with the city council, which had fallen under the influence of Maurice Hart, a businessman "with an obscure and questionable background" who controlled street paving, streetcars, and electric lights.[38] Hart's shifty practices in the city dated back to 1880, when he was involved in a bond fraud scheme.[39] It was exactly the kind of corruption that Shakspeare found so despicable with the Regulars. He noted: "I object to giving Maurice Hart or any other one man the whole City of New Orleans. He has a full power over the City Council which I do not understand. He is always asking for some privilege, and the Council gives him everything he asks for. I do not consider it right that any one man like Hart should wield the power which he does."[40] Shakspeare was confident in his chances for victory in the 1892 election, but the Ring's Fitzpatrick was elected mayor with the support of poor and working-class voters, both Black and white. Reform plans for better sanitary services fell by the wayside.

John Fitzpatrick had worked his way up to be commissioner of public works, the seat of city patronage. He used it liberally. But even with significant patronage and support from the working class, Fitzpatrick was not spared from mounting complaints about garbage service. Garbage may have been a political issue to city officials, but it was a real nuisance to city residents. In theory, garbage was collected by the city daily, taken down to the wharves, put on boats, towed downriver, and dumped, disappearing into "the devouring jaws of gar pikes, catfish, and the other greedy denizens of the great stream . . . and not a trace of any part of the offensive matters can be discovered" for miles.[41] In practice, however, the city collected garbage in a "slummiking and negligent fashion that it might almost as well be left undone." The garbage cart drivers—"mostly mere boys unfit for the work"—were not supervised, and performed only in the "most perfunctory and slipshod fashion," creating a mess.[42] The cart drivers spilled trash; their uncovered carts leaked, scattering filth in the streets and dispersing "vile and poisonous odors in the air."[43] The city's garbage collection was "shockingly defective."[44]

In 1893, the first full year of Fitzpatrick's administration, the *New Orleans Times-Democrat* railed against the inadequate garbage collection. The paper denounced past and present commissioners of public works for making "a complete and total failure of this municipal work. And, as after all these years of trial the result has been the same—failure, unmitigated failure—the hope may as well be abandoned at once that we are to have anything better with the continuation of the present unworkable system."[45]

Part of the problem may have been attributable to insufficient knowledge about the best way to deal with the city's trash. The state board of health met with city leaders and offered guidance about how other cities dealt with garbage removal: in Washington, DC, garbage was used as fill; New York dumped it in the rivers; Jacksonville and Chicago had crematories; and St. Louis used reduction by contract.[46] But the evidence suggests that the sorry state of affairs in New Orleans had less to do with knowledge than with politics. The *Daily Picayune* noted: "Why is it that the refuse is not gathered up promptly and carried away to the garbage boats? Is it because there are too few carts, or because there is official neglect?"[47] The board of health reported that "the garbage service of this city needs to be completely changed, since it is radically defective."[48]

New Orleans officials considered a number of ordinances designed to improve collection, disposal, and general handling of trash. They eventually decided to contract both collection and disposal. In August of 1893, the city council passed an ordinance directing the comptroller to advertise for bids to collect, remove, and dispose of all garbage for twenty years. The winning bid would "present to the city council a system of disposing of refuse, vegetable and animal matter, including dead animals, in a manner which is scientific and sanitary."[49] Sealed bids were collected; the lowest bid *doubled* the existing cost of collecting trash.[50] The contract went to the Southern Chemical and Fertilizer Company (believed to have been a front for Maurice Hart) to build a reduction plant touted as the largest in the country.[51] The original contract was transferred to Southern Chemical in February 1894.[52]

It is not at all clear from existing records why Fitzpatrick chose to contract garbage collection and disposal to Southern Chemical, disrupting the key source of patronage that kept his administration (and others) afloat. Speculation swirled that Hart was well connected to—or even boss of—the Democrats in New Orleans.[53] What is clear, however, is the effect this had. New Orleans's residents revolted, taking to the streets in protest. The opposition coalesced around a Mr. Maille, a lawyer and member of the city council. Southern Chemical quickly folded, and garbage collection and disposal reverted back to collection by carts and dumping in the Mississippi River.

The events that led to the downfall of the garbage contractor—the diminishing of hope that New Orleans would become a modern, sanitary city—and the end of Fitzpatrick's administration deserve closer scrutiny. They tell a story of corruption as an obstacle to building the capacity for functioning trash collection and disposal. Even before the contract was awarded, there were valid concerns that it would go to "some specially favored person, who will no more be held to a strict construction of the contract than are other specially favored contractors."[54] Newspapers recognized that the contract was a done deal, given more as a reward to the Regulars' friends than as a measure to actually clean the city, declaring in frustration that "private interests were engaged in letting out a private contract with no assurance that the service at a higher cost would be any better than at present, and no objections made by the people or the press had any influence in determining the matter."[55]

Southern Chemical and Fertilizer intended to build a reduction works, in which trash would be burned to create a sellable by-product, in this case fertilizer. In November of 1893, when Southern requested permission to locate the garbage works in the Third District, residents who lived there protested.[56] It is hardly surprising why. Sitting garbage was bad enough, but reduction emitted noxious odors to boot. Reduction also required compliance from city residents. Burning materials to produce a by-product meant that the contractor would collect only certain materials and, furthermore, what he did pick up had to be properly sorted: "He refused to cart off old shoes, bottles, broken glass, or tin cans. A special type of receptacle had to be used for garbage, or it would be left to rot on the sidewalk."[57] When burning garbage at the incineration plant was delayed, trash piled up, and the city reeked.[58]

Residents were outraged by what they saw as the city's refusal to do anything constructive about the problem. City leaders and the board of health pointed fingers over who had the responsibility to oversee the contractor. For residents, it was not just the contractor that was the problem. As the *Daily Picayune* noted:

> The persistence with which all responsibility, liability, obligation or duty has been evaded in regard to a matter which so intimately concerns the health and the welfare of the people of this city has been so remarkable that it has attracted attention, and the people had come to believe that they were wholly given into the power of a contractor who is to receive the enormous sum of $2,845,000 for services under a contract which no official authority seems willing to enforce, and under which the city government disclaims any power or function, save to pay the contractor the stipulated amounts of money upon his demand.[59]

To add insult to injury, it was the city's residents who seemed to do nothing right. In a series of escalating moves, the Fitzpatrick administration tried to induce compliance. At first, city officials recommended *educating* individuals. The mayor issued a proclamation telling all householders to use a proper receptacle for garbage, specifically a "suitable metallic, water-tight, covered box or other covered metallic vessel, in which they shall cause to be placed daily offal, garbage, slops and refuse, animal and vegetable matter from their premises, and which they shall keep in such a place as will be most convenient for the contractor to remove." The city next moved to *punitive* measures, threatening citizens who did not obey with hefty fines and even imprisonment for each day they failed to comply.[60]

New Orleans residents bristled and refused to cooperate. They were accustomed to city carts collecting garbage, loading it onto barges, and dumping it in the Mississippi River, a process that removed some of the garbage from homes. Granted, waste slipped over the sides of carts, and garbage dumped in the water washed back up onto the riverbanks and low-lying lands, but it was a process that was easy and familiar. When residents learned of the public-private contract for collection and disposal, they complained that the city had made the contract without consulting taxpayers and that the cost of the contract was too high.[61] When the mayor announced that the garbage contractor required regulation garbage boxes for daily disposal of offal, garbage, slops and refuse, and animal and vegetable matter to be placed outside dwellings between 6 a.m. and 8 a.m., with a fine of up to $10 or ten days' imprisonment for each day of noncompliance, the public was neither pleased nor compliant.[62]

Resistance to the garbage contract, Southern Chemical, and Fitzpatrick only grew. Despite the harsh penalties, many New Orleans residents didn't bother to get the required garbage boxes, which cost between $1.50 and $2.00, and few were available in the city anyway.[63] At bottom, citizens resented the fact that the city required everything from them "under pains and penalties" while the contractor felt "no pressure and no compulsion."[64] Imagine paying $3 million for inadequate garbage collection and being blamed for poor results! The old system of widows' carts and dumping in the river was no worse than Southern, and it cost half as much.

Capacity Abandoned or Lost

In both St. Louis and New Orleans, challenges to corruption sank new, sanitary forms of garbage collection and disposal, but they did so in different ways. In St. Louis the battle to rid the city of corruption meant divorcing the city's sanitation practices from Ed Butler. St. Louis, however, did not have an alternative way to dispose of trash. The nation's fourth-largest city struggled for decades before returning to dumping trash in the Mississippi River. In New Orleans, the failure of the reduction facility returned trash collection to the widows' carts and disposal to dumping on land or in the river.

Capacity Ends in St. Louis

Although Ed Butler, the conduit between the Cinch and elected officials, worked with officials without regard to party, he depended on their willingness to sell city services. Butler's lock on power changed dramatically when the reformer Rolla Wells was elected mayor. Wells was a progressive, and under his administration there were long discussions not only of health and sanitation but also of sanitary science. Sanitarians were not the only ones pushing for change. Long-simmering dissatisfaction among the public became readily apparent. Piles of uncollected, putrid trash were topped with signage calling for "15 cents Reward for a Sight of Butler, Dead or Alive." On September 4, 1902, the St. Louis Republic ran a political cartoon under the heading "Novel Methods Used to Force Collectors to Take Up Garbage." It shows three garbage pails draped in black "crape," a symbol of mourning, with a sign that reads "Where Oh, Where Is Butler?" (figure 3.2). Clearly, residents were unhappy with how infrequently Butler came to pick up trash, and their anger spilled out in signs and near fights with garbage wagon drivers.[65]

Mayor Wells was determined to take collection away from Butler not only to improve services in the city of St. Louis but also to "strik[e] at one of the deep roots of municipal corruption" and "[assert] the rights of the city over the privileges of the feudal political lord."[66] Officials, however, wrestled with how to do it. The board of health had the authority to terminate contracts for negligence, but the city didn't have many alternatives. When Wells was elected in 1901, city waste was being collected by Ed Butler's business, the Excelsior Hauling Company, under a contract for another five years. The city's contract with the Butler-friendly St. Louis Sanitary Company for disposal, however, was up in September of that year. At that time the city had the option of purchasing the reduction plant.[67] Surely this would be a legal way to wrest control over sanitation from Butler. There was one key problem: nobody in

Garbage receptacles draped in crape by householders in the vicinity of Grand and Cass avenues who believed that the garbage collector was dead.

FIGURE 3.2.   Political cartoon, "Where Oh, Where Is Butler?"
Source: *St. Louis Republic*, September 4, 1902, 14, Missouri Historical Society, St. Louis.

city government knew anything about the reduction plant or, worse yet, how to run it. According to the *Republic*:

> There is not an official in the City Hall who has the knowledge or from whom reliable information on the points can be obtained. . . . There has never been a garbage reduction works in this city till the present plant was erected, and it has at all times been operated by a private corporation. In consequence there is no general knowledge on the part of the city officials or public as

to the cost of such a plant, of the cost of its operation or the value
of its residue products.

Searching desperately for more and better information, St. Louis's
health commissioner wrote to various cities to determine what would
be a fair price to pay for garbage reduction. He received no useful guidance except from Philadelphia, which indicated that "the amount paid
by St. Louis . . . is some higher than that paid by Philadelphia."[68]

If St. Louis could not buy and run the reduction plant itself, it could
contract out to have someone else run one. This particular contract
proved a pivotal event in Butler's downfall—and the city's inability to
properly dispose of trash for years to come. The story gets very technical, but the details are revealing. On September 17, 1901, the Municipal
Assembly adopted an ordinance declaring that "the Board of Health
is authorized to provide for and enter into a contract for the sanitary
disposal of all slops, offal, garbage, vegetable matter and animal matter
in the City of St. Louis by the Merz process." Less than a week later, on
September 23, the board approved an advertisement calling for bids to
be received "at the office of the Board of Health on Tuesday, October
1st, 1901 at 4 p.m."[69] The contract was set to begin on November 17.

Six weeks is not a lot of time to set up a reduction works or even
take over the existing one. City officials certainly were not prepared
to do it themselves. It turns out, companies in the area felt the same
way: on October 1, 1901, the St. Louis Sanitary Company submitted
the only proposal, at $130,000 a year for three years, an amount twice
what it had been paid under the previous contract. To be fair, the company did propose erecting and maintaining a new reduction facility.
On Thursday, October 3, the board of health held a special closed-door
meeting to consider the garbage disposal contract. Two days later, the
board contracted with the Sanitary Company for a three-year, $390,000
arrangement.[70]

As with many other ordinances and contracts in St. Louis, it turns
out that Butler cared a great deal about who would be selected. In addition to the usual $2,500 a year that the Sanitary Company paid Butler
to promote "good feeling," it offered to pay Butler's Excelsior Hauling
$45,000 as well as $17,000 a year throughout the three-year contract if
it were awarded to the Sanitary Company. Even though the Sanitary
Company was the only bidder, Butler went to the homes of two of the
physicians on the mayor-appointed board saying, "I would like to come
to you, Doctor, and make you a present of $2,500 if you will vote for

our company to get this contract."[71] One physician, Dr. Henry Chapman, went to prosecutors. On April 5, 1902, Ed Butler was indicted on attempted bribery charges. He was convicted and sentenced to three years in prison, only to have his conviction reversed the following year.

Of course Butler tried to give money to Chapman. But it was not a crime—it was not an issue of bribery after all. The ordinance authorizing the *board of health* to contract turned out to be a key detail.[72] Even though cities were motivated to address the garbage problem to improve public health, and even though garbage collection and disposal were part and parcel of the health department's mandate, the Missouri Supreme Court ruled that garbage removal was related not to public *health* but to public *works*, which came under the jurisdiction of the Board of Public Improvements. The board of health had no right to contract for garbage removal, and therefore it was not bribery to offer money to a member of the board of health. Indeed, the mayor had signed the ordinance giving the board of health power to award garbage removal contracts a few hours *after* Butler attempted to hand the money to Dr. Chapman.[73] President Theodore Roosevelt called the reversal "an outrage" and the judges "recreant to the cause of decent government."[74]

As frustrating as the outcome of the legal case against Butler must have been, there was a faint silver lining. When the state Supreme Court decided that the board of health could not legitimately contract for garbage services, its ruling held, indirectly, that the city did not have to abide by existing garbage collection and reduction contracts. St. Louis officials were free to "put into effect any arrangements they might see fit to authorize and carry out."[75]

The city started reviewing, in earnest, its relationship with Butler. The board of health required him to provide a sworn statement as to the number of men and teams required for collection each day during the previous month, which Butler did. The board's members met with the mayor and city counselor. And they reviewed Butler's contract. "After a fair trial I was convinced that the work would never be thoroughly done until it was taken hold of by the city," said the new health commissioner.[76] The city paid an inflated price for Butler's hauling plant and took over trash collection, using "the city's own forces."[77]

Although the president of the Board of Public Improvement called attention "to the admirable manner in which the street department has collected the garbage," and the health commissioner also praised the "noticeable improvement," the "burning question" was what the city would do about disposal.[78] It was a real dilemma. As long as the Sanitary Company disposed of St. Louis's waste, Butler would retain his firm grip on the city. Mayor Wells didn't have a ready alternative.

St. Louis officials were serious about finding the right solution to the garbage problem. If they were freed from the confines of boodle—which dictated who would dispose of garbage and how—they could adopt any one in a range of solutions. Feeding garbage to farm animals hardly seemed feasible for a city as large as St. Louis; dumping garbage on land was not "in vogue in progressive cities"; and dumping offal in streams was "the worst offense of all," causing problems for both the dumping community and its neighbors downstream.[79] In a large, ambitious city like St. Louis, a better form of collection and disposal ought to have been possible.

Officials looked for scientific evidence about what worked best in other cities. The city counselor studied practices in the United States and England and determined that city-owned incineration—not reduction, which was said to be too difficult and too noxious to be done effectively—would be the best solution.[80] The health commissioner, along with a committee, went to Louisville, Montreal, Minneapolis, New York, and Washington, DC. The committee made several recommendations, including (1) enforcing primary separation, requiring householders to separate their trash into components that would be collected and disposed of in different ways (residents across the cities we examine, however, never liked to do this); (2) daily collection; (3) disposal by reduction; and (4) contracting for reduction.[81] On November 1, 1903, the commissioner recommended reduction of wet garbage by a private contractor outside the city or where there would be no annoyance to residents and incineration of dry refuse operated by the city with plants located in the north-central and south-central sections. In December, a member of the board of health reported on what he had learned about garbage disposal practices at the American Public Health Association, noting that all of the innovations in sanitation came from experimentation by eastern cities: "Your representative was not a little disappointed to find that not a single city west of the Mississippi River was making any effort whatsoever to enter into and do its share of the experimental and scientific work being conducted by this Association in the interest of public sanitation."[82]

The newly formed Civic Improvement League of St. Louis prepared ordinances providing for the collection of garbage, ashes, and refuse; for disposal by reduction plant; and for utilization of the resulting material for profit. But for all of the expert knowledge gathered and recommendations made by public health and sanitary officials, St. Louis had little time to execute any of them. City officials scrambled to figure out what to do. The city bought itself a little time when, in March 1904—a month before it opened the World's Fair—it authorized an emergency contract with the St. Louis Sanitary Company to dispose of trash until mid-November.[83] But as the Civic League noted, "The city was in imminent danger of being left without adequate service or any service at all in the matter of garbage reduction after November."[84]

Capacity Ends in New Orleans

Unlike in St. Louis, where the Cinch and Butler would work with any political official, in New Orleans corruption was deeply entrenched in the Regulars, a faction of the Democratic Party. In fact, corruption even threatened the Regulars' administration. The Citizens' Protective Association met to discuss impeachment of corrupt officials in 1894. Impeachment was the worst outcome for a corrupt politician "because [while] a pardoned convict can hold office, an impeached officer cannot." The association also wanted to strip Southern Chemical of its "illegal" contract. Apparently the state still had a law on the books requiring New Orleans to dump refuse in the Mississippi River,[85] so New Orleans's contract with Southern was technically invalid.

In 1895 Alderman Maille launched an investigation into the garbage contract and the provisions for enforcing it, calling them a "rank and monstrous fabric of incongruities" and "rotten and malodorous as the garbage itself."[86] Maille introduced a number of ordinances, asking for the forfeiture of the contract, the removal of the plant to another location in the city, and the sale of the right to remove garbage.[87] He objected mightily to penalizing citizens with fines or imprisonment for noncompliance, which "would open immense opportunities for bracing or blackmailing the poor and unsophisticated people to avoid penalties"[88]

Maille wanted to turn the tables and suspend the $10,000 monthly payment to the garbage contractor until the city attorney decided whether the garbage ordinance was constitutional. Southern was required to remove garbage in watertight, covered metallic carts but consistently failed to do so. Yet there were no penalties for Southern's noncompliance, explained away by a lack of legitimacy: "Of course, it was never intended that this law should be enforced against the garbage company, and this was shown by the refusal of the City Council to enact Mr. Maille's ordinance which proposed to inflict a fine on the contractor. . . . Nobody expected any other result from a City Council which has been publicly announced to belong to Maurice Hart, the head of the garbage company."[89] It was Hart's involvement that cast the project of garbage collection itself in a bad light: "There never was a contract for public purposes imposed upon the people of this city that redounded so little for the public good and so largely for the benefit of the contractor, and that looked so much like a corrupt job on the part of the City Council."[90] The *Daily Picayune* was skeptical: "It is not likely that Mr. Maille's efforts to protect the interests of the taxpayers and citizens from the infamous exactions of the garbage ordinance will receive any attention at the hands of City Council so entirely under the control of Maurice Hart."[91] True enough, the city council did not amend the garbage ordinances, and it did not strip Southern Chemical of its contract. But a grand jury brought nine indictments against Hart for perjury, subornation of perjury, and obtaining money under false pretenses as well as thirteen indictments against city officials, including councilmen.[92]

The reduction contract was abandoned at a great loss to the investors. Southern Chemical and Fertilizer went into receivership in 1897. More than just a temporary setback, the failure of the reduction plant set the path for New Orleans to revert to and continue dumping for another twenty-five years.[93] At the same time, the Fitzpatrick administration was engulfed in scandal: ten councilmen, the city engineer, and a former tax collector were indicted on bribery charges.[94] The Citizens' Protective Association sued Fitzpatrick to remove him from office. But Fitzpatrick won the case as well as a libel suit he brought against a newspaper. If he expected vindication on Election Day too, he was surely sorely disappointed. The Regulars were dealt a devastating blow in the election of 1896, losing fifteen of New Orleans's seventeen wards in the mayoral contest to reformer Walter Flower.

Born to a wealthy planter family, and having graduated from law school at the top of his class, Flower was everything that the corrupt Fitzpatrick was not. The *Picayune* described Flower's election as a "new era in the political conduct of this city."[95] The city put in place a new charter, with an elected mayor, comptroller, and treasurer, and a commissioner of public works appointed by the mayor.[96] Still, Flower suffered many of the same disadvantages as Fitzpatrick, and he too held the office of mayor for only one term. Although the overt controversy had ended, the sanitation problem was far from solved. The department of public works continued to collect garbage by cart in the city of New Orleans and continued to dump it in the Mississippi River.

Corruption as Obstacle

Corruption motivated political officials in St. Louis and New Orleans to take up the garbage problem. Yet corruption can also be an obstacle to policy ends. In St. Louis, ridding the city of the corrupt Ed Butler also meant ridding it of the essential capacity to dispose of trash. In New Orleans, the city never developed the capacity, and corruption ensured that it would not try again soon.

St. Louis Reverts to Dumping

St. Louis developed the capacity to dipose of trash through a modern reduction facility. It was, however, in the hands of the corrupt Ed Butler. In an attempt to clean the city of corruption, Mayor Rolla Wells inadvertently made it physically dirtier. On November 15, 1904—the day after the contract with Butler's Sanitary Company expired—the nation's fourth-largest city embarked on a "most striking departure from the manner of disposing of a city's garbage ever inaugurated in any city of the West."[97] To undermine Butler and the Sanitary Company, Mayor Wells quietly leased Chesley Island in the Mississippi River. It was meant to be a temporary experiment, and the city's lease lasted only until March 1906. But as St. Louis officials struggled to find and maintain a place for adequate garbage disposal, they resorted to using Chesley Island for more than a decade.

Steamboats would tow barges carrying garbage to the island and return three days a week, dumping two hundred tons of garbage for less than a dollar a day. The *St. Louis Republic* was jubilant, running headlines such as "President of the Board of Public Improvement Finds a Way to Dispose of Refuse at Less than Half the Amount Paid the Butlers, Which Was $356 a Day—May Even Rival [Robert Louis Stevenson's] 'Treasure Island'—Scheme is Meeting With Success."[98] Although Wells's administration improved disposal by removing the source of corruption, that meant reverting to dumping, a "primitive" mode of disposal.

The city's plan was to spread a day's worth of garbage three inches thick over each acre of sand. Men driving mules would then plow the garbage nine inches under. But Chesley Island was only four hundred acres, and using an acre a day meant that it would be little more than a year before the island was at capacity. The unlikely solution to this problem? Thousands of hogs were brought to the island to feed upon the city's garbage. The hogs consumed a portion of the refuse on the island, leaving less work for city workers (and mules). As city officials prepared short-term plans to bring in even more hogs, they also held on to long-term plans to build incineration plants. St. Louis was hopeful. Hiram Phillips received "the congratulations of the city officials for his labors in locating an emergency plant so adapted to successfully render the refuse during the time of the completion of the incinerating plants."[99]

Even with the increased garbage as a result of the many visitors to St. Louis's World's Fair, garbage collection "proved satisfactory to the public, judging from the few complaints received." It was expensive, however. St. Louis spent $194,202 to remove 65,000 tons of garbage in 1904 alone. That money paid for capital improvements—including a new wagon shop, washing plant, and office building—as well as the costs of maintaining 264 mules, fourteen horses, and 130 garbage wagons.[100]

If the city was progressing with garbage collection, it was not having a great deal of luck with disposal. The planned incineration plants never materialized. Instead, in March 1908 the city entered into a ten-year contract for garbage disposal. The contract was awarded to the Standard Reduction & Chemical Company, presided over by a supposed business associate of Mayor Wells.[101] But a year later, the company's work was judged "insufficient," the machinery "too light" to meet St. Louis's needs.[102] It was nothing short of a disaster: 5,400 tons of garbage strewn about the railroad tracks that carried trash to the plant, and the odor that accompanied it angered citizens and the Missouri state board of health. The Standard Reduction Company had a difficult time disposing of the trash delivered to it. Because the reduction company could not handle the tonnage of garbage, and because there was a $10 per ton penalty for refusing garbage brought by city collectors, Standard instead "piled it on the ground." Even if and when it expanded capacity, the company wouldn't necessarily be able to handle the accumulated garbage in addition to the current deliveries, the latter only one third the amount expected in the summer months.[103]

Once again, city officials were forced to go back to dumping on Chesley Island. The Great Western Chemical Corporation, which operated a reduction plant in Stallings, Illinois, took over disposal of St. Louis's trash. But Great Western's bankruptcy in 1913 and closure of its plant pushed St. Louis officials again to dump all garbage on Chesley Island, to be plowed under "until summer and possibly for six or eight months, until some other arrangement can be made, either through a new contract with a new company or by the erection of a municipal plant, for reduction or incineration." Chesley Island—"a foot deep with garbage" and "covered with lime"—was, in the mayor's words, "up against it."[104] Each day a portion of the garbage was dumped in the river. The city had to take action.

St. Louis officials requested bids for a six-month contract to dispose of trash. E. J. Cassily, one of Great Western's receivers, was the lowest bidder at eighty seven cents a ton—sixty cents a ton more than Great Western had charged. Yet Cassily's plan wasn't cleaner or better: he was going to take the refuse ten miles outside the city and dump it in the Mississippi River, which by this time required permission from the federal government. If a contractor could get a permit from the federal government to dump trash and make a profit, why not the City of St. Louis? City officials hatched a secret plan, and when they were assured the federal government would allow it, they made the plan public. Despite interest from civic associations and careful study and recommendations by health officials, the city failed when it came full circle, dumping trash in the Mississippi River as it had done in the 1870s, before Butler became involved.[105]

New Orleans Reverts to Dumping

As in St. Louis, where lost capacity had long-term effects on trash disposal, New Orleans's failure, too, had long-term effects on sanitary practices. Although the Ring was wounded in the election of 1896—losing the mayoralty to reformer Walter Flower—it regained its strength. The *Daily Picayune* had cautioned that "the ring snake was scorched, without being killed." In other words, the crushing electoral defeat at the hands of Flower stung the Regulars, but it did not stop them. In fact, it gave them the opportunity to rethink the way they operated. The Regulars modeled themselves on the better-organized and more successful Democratic political machines. They reconstituted themselves as the "Choctaw Club of Louisiana," a nod to the Iroquois Club, Chicago's machine. John Fitzpatrick was the club's new leader, and he counted Governor Murphy Foster as a keen supporter, providing extensive patronage in state jobs to club members.[106]

Fitzpatrick and the club pushed for disenfranchisement of the state's Black population in 1898. They claimed that the reformers had manipulated the ballots of Black voters, leading to the Democrats' defeat. Fitzpatrick and other members of the club encouraged voters in the Third Ward to withdraw suffrage from the "worthless class." When the Constitutional Convention met in February of 1898, club Democrats occupied key positions, including president of the convention. Fitzpatrick was a powerful delegate, assigned to the committee on organization and rules, the committee on New Orleans, and the committee on suffrage and elections. These positions allowed him to push for disenfranchising Black voters at the same time he made sure that immigrant voters—particularly those who could not read or speak English, who were a key base of support—were protected.[107]

The convention marked the turning point for New Orleans's Democratic Party. By denying Black residents the vote, protecting immigrant rights, and providing patronage at both the state and city level, Fitzpatrick had engineered a long-term solution to the instability of the city's politics: a political machine with muscle. Voters embraced the machine in the election of 1900, and the new mayor thanked Fitzpatrick for his victory. The reformers folded. Fitzpatrick was elected to the state legislature. But his real legacy was in stabilizing New Orleans's politics, ensuring that the Democrats would rule the city for decades.

In 1904 the machine's Martin Behrman was elected and served four terms. He was narrowly defeated by Andrew McShane in 1920 by only seven hundred votes. The Behrman administration was known for its civic advancements and substantial improvements.[108] It tried some innovations, shifting from dumping garbage in the river to burying it in outlying areas. The city laid out and enforced rules for carts and drivers, requiring proper uniforms; specifying cart dimensions, cart signage, and the condition of the animals; and admonishing drivers to be polite (for example, no foul language). In addition, drivers had to be over eighteen and sufficiently strong. They also had to be neat, and to wear a red shirt.[109]

Garbage collection was imperfect but seemed to stabilize after the contracting-out debacle. The city employed an average of 110 carts; forty thousand out of fifty thousand homes had daily cart service; and the department of public works recorded as few as eight complaints a month.[110] The system of widows' carts continued. In January 1908, George Smith, commissioner of public works, informed the superintendent, "You are hereby requested to place the following carts at work," listing the names of six owners, five of whom were women.[111] Even as the patronage system persisted, Smith sought to enforce personnel rules and pleaded with the city for resources to buy garbage boats. The existing boats were in poor shape and eventually sank. The result was that "tidal overflows immediately reincarnated the all too substantial ghosts of last week's cabbage and yesterday's shrimp heads." Burning the garbage (in piles, not in an incinerator) created fumes. Dumps invited "hordes of rats, mice, flies, mosquitoes, and buzzards."[112] The city's Progressive Union, "guided entirely by a sense of public duty, and an earnest and commendable desire to assist in removing obstacles and impediments to a clean city," provided the text for a new ordinance that would prohibit garbage from being put on the sidewalks, and it made Mayor Behrman a director of the union.[113]

Correspondence from the department of public works indicates that the Behrman regime's civic improvements were not improving the garbage situation. Commissioner Smith had spent years reporting on the precarity of the garbage boats and wharves, letting the city council know of the poor conditions on board and warning that they required "constant attention in order to keep them afloat."[114] His pleas went unanswered until he described a boat sinking at the wharf, despite "every means within the power of this department . . . employed to endeavor to raise this boat, but to no avail."[115] He eventually got the garbage boat situation under control, touting his department's resourcefulness in using wastepaper and trash to fill an old canal which had been both an eyesore and a breeding ground for mosquitoes.[116] The Behrman administration may have promised progressive reform, but Smith continued to deal with garbage collectors arriving late for work and foremen failing to enforce the rules that Smith so assiduously disseminated, even having to remind foremen that workers should not be intoxicated or attend baseball games during working hours.[117] Those behind-the-scenes troubles were quite evident publicly. In 1918 the Behrman administration oversaw the erection and use of a new incinerator. But even as late as 1926, the city struggled with incineration.[118]

Failures in Capacity

Corruption in St. Louis and New Orleans provided the resources that would move public officials to act on the sanitary concerns raised by the experts. Garbage collection promised a lucrative contract for St. Louis's boodlers. The need for a team of collectors and cart drivers could be met by the patronage-by-ward system practiced in New Orleans. The need for a fleet of carts could provide charity to the widows of party loyalists, too. If the sanitarians and related experts were unable to actually pick up the garbage themselves, the variety of corrupt practices already in place in these cities opportunistically provided the incentive to act on the sanitarians' warnings.

Establishing a muncipal collection program did not mean that garbage was actually picked up, however. Both St. Louis and New Orleans contunued to be plagued by filth, whether from trash not being picked up or tumbling out of carts in transport, or through poorly designed disposal practices. Corruption provided the incentive for municipal garbage collection but not necessarily the capacity, as St. Louis and New Orleans vividly illustrate. Although corruption in these two cities prevented development of the capacity to collect and dispose of trash, in Charleston and Pittsburgh (chapter 4), corruption furnished both political will and capacity.

CHAPTER 4

Picking Up Trash

Adequate Garbage Collection by Corrupt Regimes

Like St. Louis and New Orleans, both Charleston and Pittsburgh had a garbage problem that prompted trash collection and disposal programs. Charleston took the initiative in 1806, empowering the commissioners of streets and lamps to remove trash, but Pittsburgh did not allow for the systematic removal and disposal of trash (and here by contract) until 1895. Both cities, however, created functioning (albeit imperfect) trash collection and disposal programs. Why did Charleston and Pittsburgh create programs nearly ninety years apart? And why were these programs adequate when those in St. Louis and New Orleans were not?

In this chapter we show how corruption fostered solutions to pressing policy problems in Pittsburgh and Charleston. As in St. Louis and New Orleans, corruption motivated city officials, who saw garbage as another means to benefit themselves, to create collection and disposal services. Unlike in St. Louis and New Orleans, corruption in Charleston and Pittsburgh was integrated into government, and it gave city officials the capacity to put trash collection programs in place: Charleston elites relied on their position in government and connections to exploitative slave labor and its legacy, while Pittsburgh was able to use the political power, financial resources, and business acumen of the machine.

Corruption served as a motivation and resource for both governments. It allowed Charleston's elites, who ruled in the early nineteenth century, to make money by providing a new service, leasing the labor of enslaved persons. Charleston used the garbage as fill, expanding the city into the marshy lowlands. When Pittsburgh's Republican machine finally came to power, corruption provided both motivation and capacity to pass a state law and a city ordinance enabling the city to collect and dispose of trash as well as to contract out to machine affiliates to build a state-of-the-art disposal facility. Even though Pittsburgh's health officers had argued for decades that cholera and typhoid outbreaks were caused by unsanitary conditions and pushed for better sanitation, the city was unable to do anything more than use nuisance laws until the machine came to power in 1895.

Corruption, however, did not look the same or function the same way in the two cities. Charleston was ruled by a set of leading families that formed a complacent quasi-aristocracy, while Pittsburgh was run by a political machine. Corrupt regimes had access to different resources, such as family connections in Charleston and money in Pittsburgh. Yet in both cities corruption was integrated into government, so there was little distinction between what the corrupt regime wanted and what the city pursued and how it did so.

In the case of garbage, corrupt interests and corrupt actors in both cities were motivated to collect trash not because they wanted good, effective public goods but because they could enrich themselves by providing this new service. Charleston's wealthy families were employed in government, and they benefited both from their own employment and from hiring out their slaves to work for the city. Pittsburgh's machine was able to exploit the garbage problem to enrich itself by contracting out to machine actors to provide garbage collection and disposal.

Far from hindering a new service as in St. Louis and New Orleans, corruption was a resource in Charleston and Pittsburgh. Both cities gained the capacity of their corrupt regimes to address trash: Charleston gained the elite's capacity to exploit slavery, while Pittsburgh gained the machine's business capacity and resources. In what follows, we detail how corruption motivated political officials in Charleston and Pittsburgh to take up garbage collection and disposal and how it also provided the capacity to do so. We show that *when* these two cities chose to provide garbage collection and *how* they chose to do so were directly related to the type of corrupt regime that ruled each city, the ways each regime could benefit by providing this new service, and the capacity each had to get the job done.

The Integrated Regimes of Charleston and Pittsburgh

Charleston and Pittsburgh were dirty places. The board of health lamented Pittsburgh's reputation as "filthiest city of the Union," but descriptions of Charleston suggest it was a contender for claim to the title.[1] The *Charleston Daily News* presented the garbage problem vividly:

> The filthy accumulations of a city composed of all the refuse matters that collect after the wants of a people have been satisfied—the remnants of butchers' meat, the parings of vegetables, the pluckings of poultry, the washing of culinary and domestic utensils, the decayed fruit, the spoiled fish, the bones, feathers, scales and other refuse parts of cooked and uncooked animals, birds and fish, the carcasses of domestic animals, the sour remnants of the table and kitchen, the greasy and fetid water in which these remnants have been fermenting—in short, all that collects from the satiated wants of man and animals, the excrementitious matters from the stables and cowhouses, all these and an endless list of other impurities which mix with them and make up the foul burthen of the odious scavenger's cart, go forth daily to be cast into the low, moist and muddy places of the city, to foment and putrify till they saturate the air with filthy emanations, pregnant with disease and death.[2]

The politics were no cleaner. Late-nineteenth-century Charleston was run by its quasi-aristocracy, wealthy elites who had effectively ruled the city for years, while Pittsburgh was headed by a political machine, an organization "interested less in political principle than in securing and retaining political office for its leaders and distributing income to those who run it and work for it."[3] Members of Charleston's aristocracy and Pittsburgh's machine used their positions in government to enrich themselves, the aristocracy through nepotism and the machine through rigged contracts. The story of garbage collection and disposal in both cities shows how formal democratic structures can use corruption to address public problems to benefit themselves while at the same time they develop workable policy solutions.

Charleston's Quasi-aristocracy

Even within South Carolina, which is known for sustaining an aristocratic regime, Charleston stands apart. Charleston was prosperous, enjoying the mercantile profits of the slave-based planter economy, and this wealth was concentrated in a set of old families.[4] Charleston's wealthy families in agriculture—and later trade, medicine, and law—ruled the city, dominating elected and appointed positions through nepotism, defined as the "bestowal of patronage by reason of ascriptive relationship rather than merit."[5] While the aristocracy and nepotism do not account for some incursions by newcomers, South Carolina, and Charleston in particular, showed a commitment to the old guard and a resistance to change.[6]

Tom Shick and Don Doyle illustrate this tendency with their account of the South Carolina phosphate boom. Used to manufacture fertilizer, phosphate was discovered in 1860, but the industry did not get underway until after the Civil War. A new mining method required innovations in engineering, investment, and labor. Shick and Doyle find that local capitalists were reluctant to invest. Eventually there was a mining industry, but it petered out by the turn of the century.[7] The underutilization of this natural resource is curious, given that Charleston's shipping industry was suffering at this time. Having never recovered from the Panic of 1819, Charleston fell from the rank of sixth-largest city in the United States in 1830 to twenty-sixth in 1870, and down to ninety-first in 1910.[8] Its shallow harbor was not deep enough for the new ships, and the railroad diverted South Carolina's crops from the port.[9] One might presume that business elites would have reinvested in new technologies and industries, but instead, Charleston's elite showed resistance to the change and progress that were being enjoyed in other US cities. While other factors might explain the limited success of the phosphate boom, such as reluctance of Black labor to engage and Charleston's exclusion from Governor Benjamin Tillman's regime, Doyle suggests that "we may better understand Charleston's relative decline, not as the failure of frustrated bourgeoisie to promote their city, but as a quiet triumph of a conservative local aristocracy to preserve their city and their social status within it."[10] Even with the end of slavery and the decline of the local shipping industry, the old guard in Charleston was able to retain its hold on power through its wealth, exclusive associations, and networks.

This old guard was not entirely resistant to improvements, and Charleston did accomplish some development in city services and public projects.[11] Nevertheless, aristocracic complacency explains certain well-worn paths. Streets in Charleston were notoriously unpaved. Some were filled in with cobblestones from the ballast of ships, but otherwise street paving was a job that the city offices talked about more than they accomplished.

Elite families exercised disproportionate influence in South Carolina generally, and in Charleston in particular. Of the state's 440 largest slaveholders in 1860, 41 percent served in federal, state, or local offices.[12] In Charleston generations of the same families served regularly as commissioners on the boards of health, schools, parks, city planning and zoning, port utilities, sewage, cultural institutions, and public charities, serving also as aldermen and sometimes as mayors, recorders, or police chiefs.[13]

The city's wealthy families made money on city contracts, a privilege that rested on their family ties and on their status as slave owners. Charleston relied extensively on slavery to provide a host of city services. Slaveholders "hired out" slaves to the city for public works projects (like well digging and street cleaning) as well as for public safety (in the fire department and sheriff's office). In July 1849, for example, Robert Yates, the city sheriff, was paid $12 for one hire, while another Yates, J. D., was paid $24 for the hire of two for a month. Hiring enslaved persons cost significantly less than the labor of white workers, who records show were paid $1 a day. According to one estimate, 23 percent of the city's 19,500 enslaved persons were hired out.[14] After the Civil War and Reconstruction destroyed the relationship between Charleston's aristocratic families, slaves, and the city, these families and city leaders found news ways to reinstate an old racialized hierarchical order.

Pittsburgh's Machine

Pittsburgh's ruling regime, by contrast, did not have the long familial history of Charleston's. The city was run by a relatively young machine, a partnership between city treasurer Christopher Magee and Republican Party chairman/businessman William Flinn that got underway in the 1870s. Magee amassed power through the regulation of franchises, deposit of funds, and awarding of contracts.[15] The machine controlled the city council, mayor's office, and department of public safety, which oversaw the city's trash efforts.

Magee and Flinn were both businessmen, in transportation and street paving, respectively. They maintained a tight connection between city government and local businesses, which—as with many political machines—boosted their electoral advantage.[16] Magee and Flinn ran for local and state offices, but their real influence was not in the official, elected positions they held but rather as the unelected bosses of the machine, where they worked to install their candidates in prominent offices. The *Pittsburgh Post* sardonically recounted what it looked like when the machine supported a political candidate:

> The policemen, firemen, street sweepers, sewer cleaners, elevator men, electricians, clerks, bookkeepers, park laborers and the several thousand people constituting the payroll will be personally informed that [William J.] Diehl is the choice of Senator Flinn, and that they had better see to it that their fathers, brothers, nephews and neighbors 'are right,' and that they get out to talk to their friends and get them to the primaries to vote for Diehl. This will be quickly followed up with like notice to the employes [*sic*] of the many corporations owned and controlled by the machine leaders and by the heads of corporations the machine leaders control. This will quickly bring all the men of the Booth & Flinn company, which paves all of the city streets, and all of those of the Consolidated Traction Company, which holds the most valuable city franchises, under orders.[17]

Although Charleston's old aristocratic ties and Pittsburgh's relatively new machine were organized differently and favored different constituencies, they were both corrupt, and that corruption went hand in hand with city government.

How Integrated Regimes Work

In Charleston and Pittsburgh, formal, democratic institutions were integrated, or intertwined, with informal, corrupt ones. In integrated city governments, there was broad overlap between what the corrupt regime wanted and what the city prioritized as well as the resources corrupt regimes had access to and the way cities went about implementing policies. Interests in both of these cities used their political influence often to make their actions—although entirely unethical—legal, and the city's formal capacity to provide services fundamentally relied on the informal regimes' capacity.

Charleston's elite social order directly benefited from garbage services because the governing arrangements served its interests. In the years before the Civil War, superintendents of streets and inspectors for the street department during this period came from some of the city's leading families. The list of city officials is populated by familiar Charleston names—Chisolm, Pinckney, Clement, Seignious.[18] Contracts were likewise awarded to family members and others from the inner circle. For example, the brother-in-law of Mayor John Schnierle, Henry Horlbeck (relative of Dr. Henry Horlbeck), had numerous contracts with the street department.[19] Even after the end of slavery, the old guard retained its power through its wealth, exclusive associations, and networks. They were well represented in state government but were represented even better in municipal government.

Pittsburgh's machine, too, was able to redirect democratic processes to further its own interests, tightening its grip on the city by changing the city charter in 1887. It took the power of appointments away from the city council and granted it to department heads. The *Pittsburgh Post* called it the "Mageesburgh Charter," and pronounced that Pittsburgh had been "given over to the boss." For Magee and Flinn, the new charter and a law known as the Upperman bill, which authorized cities of the second class, such as Pittsburgh, to provide for improvement of streets, lanes, alleys, highways, sewers, and sidewalks, and required "plans" for streets, "set the table for a veritable banquet of lucrative public works."[20] Muckraker Lincoln Steffens described the effect: "The Magee–Flinn machine, perfect before, was made self-perpetuating. I know of nothing like it any other city. Tammany in comparison is a plaything, and in the management of a city [New York City's Boss Richard] Croker was a child beside Chris Magee."[21]

The state of Pennsylvania and city of Pittsburgh authorized public works projects by contract, in theory to secure the best services at the lowest price for the city. But the machine manipulated contracts in its favor. E. M. Bigelow, who was Magee's cousin, oversaw the department of public works, a haven for corruption. Bigelow could reward machine businesses that bid on city contracts. As it turned out, many city projects— from paving roads to installing sewers—required ripping up the streets and re-grading, paving, and curbing. It was a potential boon to street paving contractors, but one in particular seemed to benefit from a disporportionate share of contracts: Booth & Flinn, Ltd.[22] As the head of public works, Bigelow wrote city contracts to require a very specific type of asphalt from a pitch lake on the island of Trinidad, which was conveniently owned and controlled by Booth & Flinn.[23] The Booth & Flinn street paving company dominated Pittsburgh's paving contracts (figure 4.1).[24]

Despite the many differences between the two cities, they shared with each other (and with other corrupt cities) the understanding that problems—like trash—could be addressed with corrupt resources and that these problems and their solutions could benefit corrupt leaders.

LETTING PLANS MADE AND CONTRACTS AWARDED—Continued

NAME.	FROM	TO	CHARACTER OF WORK.	MATERIAL.	CONTRACTORS.
Main street,	Present Pavement,	Butler street,	Repaving,	Block stone	Booth & Flinn, Ltd.
Miltenberger street,	Fifth avenue,	South street,	Repaving,	Block stone	Booth & Flinn, Ltd.
Monongahela street,	W. J. Lewis property,	Hazelwood avenue,	G., P. & C.,Asphalt,		Booth & Flinn, Ltd.
Murray Hill avenue,	Fifth avenue,	Wilkins avenue,	G., P. & C.,Block stone		Booth & Flinn, Ltd.
Millvale avenue,	Centre avenue,	Bridge over Penn'a R. R.	G., P. & C.,Asphalt,		Penn'a Asphalt Paving Co.
McClure avenue,	Shady avenue,	Beechwood avenue,	Grading,		Booth & Flinn, Ltd.
McCully street,	Highland avenue,	Heberton street,	G., P. & C.,Asphalt,		Booth & Flinn, Ltd.
McCandless street,	Stanton avenue,	Butler street,	G., P. & C.,Block stone		Booth & Flinn, Ltd.
McCully street.	Negley avenue,	Haights avenue,	Sewer,	Pipe,	Hughes & Meisold.
Nineteenth street,	Penn avenue,	A. V. R. R.	Repaving,	Block stone	Evan Jones.
Niagara street,	Craft avenue,	Carolina street,	G., P. & C.,Asphalt,		Booth & Flinn, Ltd.
Pocussett street,	Beechwood avenue,	Schenley Park,	Grading,		Thomas McNally.
Pike street,	Fourteenth street,	Sixteenth street,	Repaving,	Block stone	Booth & Flinn, Ltd.
Roberts street,	Enoch street,	Linton street,	Repaving,	Block stone	Booth & Flinn, Ltd.
Ross street,	Diamond street,	South street,	Repaving,	Block stone	Booth & Flinn, Ltd.
Ridge avenue,	Hancock street,	Thirty-third street,	Sewer,	Pipe,	Keeling & Ridge.
Stanton avenue,	Euclid avenue,	Haights avenue,	Sewer,	Pipe,	Sweeny & Houston.
Sarah street,	South 17th street,	South 18th street,	Sewer,	Pipe,	J. M. Deiterle.
Sicklesst.&Karlav, et al.	Singers property,	Tioga street,	Sewer,	Pipe,	Keeling & Ridge.
South 14th street,	Breed street,	Sarah street,	Sewer,	Pipe,	J. M. Deiterle.
St. Andrews street,	Omega street,	River street,	Sewer,	Pipe,	Keeling & Ridge.
Sciota street,	Atlantic avenue;	Osceola street,	Sewer,	Pipe,	T. H. Rose.
Stanton avenue.	Bet. Heberton st. and Wellsley avenue,	Highland avenue,	Sewer,	Pipe,	Evan Jones.
South 14th street,	Muriel street,	Breed street,	Repaving,	Block stone	Booth & Flinn, Ltd.
South 22nd street,	Carson street,	Wharton street,	Repaving,	Block stone	Booth & Flinn, Ltd.
Sarah street,	South Tenth street,	East,	Repaving,	Block stone	Booth & Flinn, Ltd.
South 18th street,	East Carson street,	South,	Repaving,	Block stone	Booth & Flinn, Ltd.
Soho street,	Centre avenue,	South,	Repaving,	Block stone	Booth & Flinn, Ltd.
Spring alley,	Seventeenth street.	East,	Repaving,	Block stone	Booth & Flinn, Ltd.

DEPARTMENT OF PUBLIC WORKS

57

FIGURE 4.1. Booth & Flinn receive disproportionate share of paving contracts
Source: "Letting Plans Made and Contracts Awarded," *Annual Report of the Department of Public Works of the City of Pittsburgh for the Fiscal Year 1897* (Pittsburgh: Herald, 1898), 57, Detre Library and Archives, Senator John Heinz History Center, Historical Society of Western Pennsylvania, Pittsburgh.

But in these two integrated cities, "the state [was] institutionalized to serve their purposes."[25] Charleston's hiring out system benefited its elites. *When* Charleston chose to address the problem of garbage and *how* it chose to do so are a direct reflection of the interests and resources of the informal regime. Similarly, in Pittsburgh, the machine-owned businesses could profit from public works—like clean water and sewers—and machine politicians were eager to pass legislation authorizing a host of projects. Pittsburgh pursued a number of city-led innovations, including garbage collection and disposal. *That* Pittsburgh chose to address sanitation and *how* it chose to do so is, as in Charleston, a direct reflection of the ruling regime.

Corruption Generated Political Will for Garbage Collection

Charleston and Pittsburgh developed very different types of garbage collection at very different points in time. But corruption generated the political will in both cities to do something about the trash problem.

Charleston Antebellum Collection

Charleston was an early adopter of garbage collection and disposal services. In fact, as early as 1806 it provided one of the nation's first garbage collection programs.[26] The purpose of this collection was not primarily cleaning up refuse; it was collecting trash to be used as fill for the downtown area. Charleston collected garbage to build up land in the marshes, or any area that needed shoring up, such as the Poor House wall. Rather than being treated as a problem, garbage was an opportunity, with scavengers tasked in 1806 to gather rubbish to extend the boundaries of the city.[27] City scavengers who cleaned out the gutters and filled holes in the streets were authorized to employ "three negroes or other servants" to remove rubbush from the streets and two more to level and rake.[28] Enslaved persons served as an inexpensive source of labor that could in turn serve as a resource of government.

The city built up a nascent administration as well. Charleston's official city scavengers, under the purview of the commissioners of streets and lamps, were responsible for collecting trash and disposing of it where the commissioners determined. In addition to enslaved persons or servants, each city scavenger was to employ "three horses and three carts, for the purpose of removing from every street, lane, alley . . . all such dirt rubbish . . . and likewise two negroes or other servants, provided with wheel barrows and other appropriate implements, for the purpose of levelling and raking the streets, and of keeping the gutters and grates of the drains in his division open and free from sand, filth and other obstructions." The drivers of the scavenger's cart would ring a "hand-bell opposite to every house," giving notice to the dwellers that the trash should be brought out. Carts were "painted red, numbered and marked with the words '*Scavenger's Cart*,' exhibited in large white characters on both sides of each cart." There were penalties for any person who hired or employed the scavenger's cart, horse, or laborers for work that diverted them from the public work.[29]

These hired-out persons were disciplined through rules of behavior and dress. After complaints that scavengers failed to make use of their bells, drivers were required to carry large handbells to announce their arrival. City scavengers, who supervised this operation on the ground, were accountable for the condition of horses and carts, as well as for the workers. They were regulated, in turn, by the Board of Commissioners of Streets and Lamps. The board made sure that the city scavengers dumped gathered garbage into the streets as fill, rather than make a deal with a private property owner to fill an individual's own low-lying yard.[30]

In 1858 the city passed an ordinance abolishing the office of the Superintendents of Streets, and the city council had discretion to contract out the work to the lowest bidder. The ordinance further specified that garbage should be collected daily (except Sundays), and also specified the time by which each resident should "have the dirt, filth, garbage or other offal, placed in front of his or her lot, in a barrel, box, or heap, and in readiness for the contractor."[31] This experiment with contracting would be brief. As cities across the nation established municipal garbage collection in the latter part of the nineteenth century, Charleston already had the tools in place.

Post–Civil War

During the federal occupation of Charleston after the Civil War, with the administration of military-appointed mayors, the city stopped most unessential work. The street department, the most costly department after the police, eliminated all services except scavenging. Despite limited resources, the city council notified residents to have their "dirt, filth, garbage or other offal" placed in front of their lot "in a barrel, box, or heap" to be ready for the scavengers, subject to a fine for non-compliance.[32] According to the detail of expenses, which amounted to $17,640.88 for one month (June 1867), $10,000 was for carting and $842.16 for scavengers.[33] Charleston's residents were unhappy with the sanitary conditions of the city and with the contracted scavengers' work, complaining of carts piled too high, spilling garbage onto the street in "respectable" parts of town.[34]

To gain more control over the scavenger services, and despite the economic climate, during the administration of the military appointee Mayor Milton Cogswell (1868) the city passed an ordinance for the better regulation of the street department, which made street inspectors responsible for reporting to the mayor "all neglect of scavengers, grievances and nuisances which may come under [the inspector's] observations." Furthermore, the street inspector was to keep a daily record of all laborers, carts, and materials used, including the names of the cart owners. Additionally, the inspectors were to keep a "complaint book," including citizen complaints as well as police reports.[35] In a council meeting in April 1868, Mayor Cogswell suggested that the council buy its own carts and horses to do scavenger work,[36] an indication that contracting out the service might not have been working. Mayor Cogswell's report to the city council in July 1868 detailed the "neglected condition of the streets, drains, gutters and sidewalks."[37]

Reconstruction

In 1868 Gilbert Pillsbury, a New England abolitionist, won a contested election for mayor.[38] He was committed to fostering the liberty of newly freed slaves.[39] Pillsbury's tenure was plagued by charges of nepotism and resentment at federal occupation. J. H. Jenks—who was elected city inspector and was Pillsbury's stepson—was in charge of overseeing garbage collection and disposal, which were still at this time contracted out.[40] During Jenks's term as city inspector, the city's sanitary condition declined, and Jenks was denounced as negligent, with calls for his resignation. Overall, Pillsbury's administration was condemned for corruption and overspending. But the corruption charges, specifically the patronage charges, were also fueled by racist sentiment, as more Black workers were recruited into the municipal force,[41] including hires made by the street inspector.

In his report for 1870, Inspector Jenks writes about the sanitary condition of the city when he assumed office, blaming the previous administration in noting that "the slops and filth of entire blocks are discharged upon the surface gutters of the street, there to ferment, evaporate and yield their nauseous gases to the disgust of all passers-by."[42] It was not clear that sanitary regulations were actually enforced. Newspapers, however, were frank about the public discontent with the city inspector and scavengers, whose job it was "to see that all garbage, ashes, and other refuse matter is thoroughly removed from the city streets before 10 A.M. each day . . . but owing to the negligence of the scavengers this refuse matter is too often left on the street until midday, to fester and decompose in the hot sun." The *Charleston Daily News* condemned this "reckless and criminal disregard" of duty.[43]

Newspapers were also indignant about Jenks's salary, $12 a day: "Suppose this little job to be given out by contract to the highest bidder. Does anybody believe it would cost the city anything like that figure? Men would be glad to do the work for less than half the money!"[44] Furthermore, Inspector Jenks had been absent from his duties (somewhere in the "North") for the entire month of August 1871. The volunteer assistants to the board of health of Ward 4 asked for Inspector Jenks's suspension, given the spread of yellow fever.[45]

By the 1870s, the sanitary condition of Charleston was not too different from conditions in the late 1850s, which was due in part to "the disorganized condition of public affairs caused by the war."[46] During Johann Wagener's tenure as mayor, the city took over direct supervision of scavenger services.[47] But city oversight was no improvement. Although there was consensus among the medical establishment regarding the danger that dumping garbage in town posed to residents' health, city leaders were ambivalent and focused on the costs of disposing of garbage beyond the city limits. For example, in 1873, during a discussion of a bill prohibiting dumping garbage in lots or streets, some councilmen emphasized the expense of transporting garbage out of the city. The legislature decided that offensive material should be dumped at the public cemetery away from the downtown residential and business district.[48] Evidence suggests that convicts were used to manage the garbage deposited in the cemetery lands from at least 1871, when prisoners were assigned to maintain the Public Cemetery grounds.[49]

Charleston had limited options because it had enormous financial obligations. A significant portion of the city's budget in 1873 was destined for interest payments on the public debt ($290,000 of the $660,000 proposed budget). With the remaining resources, the city appropriated $45,000 for "streets, pavements and scavengering" and $20,000 for the Board of Health.[50] Still, city leaders did not rein in spending enough. The Wagener administration overspent by $258,189, eroding support for the Democratic Party and boosting the electoral chances of the Republicans. George I. Cunningham, a Republican, was elected to office in 1873. To address the debt incurred by Democrats but also the "huge bonded debt" from the antebellum period which almost equaled the state of South Carolina's public debt—the Cunningham administration increased taxes, lowered costs, and "issued bonds to redeem past-due city stock." Cunningham was credited with improving the fiscal health of the city during the 1870s. Among the cost-saving measures was the reorganization of the board of health and the emphasis on the cleanliness of the city to prevent costly epidemics. Yet sanitary conditions remained poor. In 1875 the city registrar, Dr. Robert Lebby, described street drains as clogged and "little more than 'Lines of cesspools' that received the 'sewerages of houses and privies.'"[51]

After Reconstruction

After 1876 the old political order reasserted itself, and by 1895 Black men were effectively disenfranchised in South Carolina.[52] Through a combination of violence and intimidation, integrating immigrants and their descendants into a white supremacist ideology, and preventing Black Americans from voting, the former means of access and inclusion were reinstated. Nonetheless, cleavages were reemerging between the white working classes and the Broad Street Ring, composed of "prominent attorneys and businessmen who controlled the city's politics."[53] Referred to as "the aristocrats," they supported "blue-blood" mayors, whose hold over city politics attests to the continuity of Charleston's aristocratic/elite rule.[54] According to Walter Fraser, the leading business interests prior to the Civil War resumed local leadership afterward. Unlike the "new" men who pursued economic development in other southern cities, Charleston's leading names were the familiar names of old—De Saussure, Gibbes, Huger, Middleton, Pinckney, Ravenel, and Rhett. They did not pursue new programs or welcome newcomers. As Fraser wrote, "A common saying put it that in Boston a man was assessed on the basis of his intellect, in New York on how much money he had, and in Charleston on 'who was his grandfather.'"[55]

In the 1880s the city increased the number of wards from eight to twelve, with two elected aldermen from each, and extended mayoral and aldermanic terms from two to four years, helping secure conservative Democrats' control over city politics. These changes, combined with the virtual disenfranchisement of Black voters in 1895, turned Charleston politics into single-party rule. By this time, institutional segregation had gained momentum, with everything from separate asylums to separate cemeteries for Black and white people. The Black political class all but disappeared. By 1883 the city council was composed exclusively of white Democratic Party members.[56]

The Broad Street Ring's aristocratic connections were evident in the appointments to important positions, including health officers, street superintendents, board of health members, superintendents of carts, and sanitary inspectors. For example, the street department was under T. A. Huguenin, son of a wealthy planter, for two decades, and he was succeeded upon his death by another member of the city's elite, James B. Keckeley.[57] The superintendent of streets, Captain J. D. Jervey, hailed from an old Charleston family, while the board of health consistently throughout the period included representatives of the leading families, featuring names such as Middleton, Grimké, Moise, Ravenel, Smyth, Huger, and Buist.

During the late 1870s and into the early 1900s, there was continuity and stability in the administration of garbage policy, in part because of the uninterrupted leadership of the street department and the health department by Thomas A. Huguenin and Dr. Henry Horlbeck, respectively. Garbage disposal continued to be completely under the control of the city, which was restated over and over in annual reports throughout the period. City collection was touted as advantageous compared to letting it out by contract. According to the city's 1895 *Year Book*:

> The Street Department removes all the garbage and refuse from the streets of the City—owns the carts and controls them, hires the drivers and controls them, and has a superintendent of carts, whose business it is to overlook and supervise these carts and see to it that the garbage is removed at an early hour in the day. The system has worked exceedingly well. The garbage is removed with certainty every day in the year except Sundays and Christmas Day. With this method there is no doubt about the service; there is no contractor to neglect his duties. It is a sure thing, and it is a great boon to the City to have the streets clean. It is the duty of the Sanitary Inspectors to report any negligence that may occur, and this is rarely required.[58]

In the early 1880s, garbage removal was placed under the superintendence of the board of health, sharing responsibility with the street department, and remained there until 1904, when, during the administration of Mayor R. Goodwyn Rhett, the garbage collection department was placed under the newly created board of public works. Garbage was collected three or four times a week, depending on the time of the year, instead of on a daily basis. Efforts focused on eliciting citizen compliance. Garbage disposal continued to be the biggest challenge in garbage administration, and the lack of a sewerage system represented the biggest sanitary concern of the time.

Pittsburgh

In the early nineteenth century, at roughly the same time as Charleston, Pittsburgh also took action. Unlike Charleston, however, Pittsburgh did not organize citywide collection. Instead it relied on nuisance laws, fining individuals $10 to $20 for dumping in streets or canals.[59]

Pittsburgh's health officials had argued for decades that the city needed to do more to address sanitation effectively. Yet the response—as in many other cities—was only sporadic. In 1832, after a cholera epidemic, Pittsburgh devoted resources to street cleaning and garbage removal. In 1851 the city created a board of health, empowering it to remove items that imperiled public health—but largely with the authority to quarantine and respond to nuisances. An 1859 nuisance law prohibited dumping trash in public places or in one's yard. An 1872 ordinance allowed the board of health to employ scavengers as needed. Those who did not comply with dumping ordinances were fined $5 for each and every day garbage remained on the premises. Those who carted away dead animals or grease in uncovered wagons were also fined.[60]

When a typhoid outbreak hit the South Side neighborhood—an area that was regularly beset with epidemic disease—board of health officials suspected impure water as the cause. They considered that the source might be the local water supply, but they also suspected that disease could be spreading because of the condition of the homes. Frame houses on the South Side were built without cellars, allowing filth to pile up underneath. The ground under the houses was "soft and boggy as ashes, filth, and garbage were used as fillings."[61]

Sanitarians' recommendations were largely ignored. In addition to dirty streets and water, the city suffered through continued disease outbreaks. Despite the need and the pleas, Pittsburgh did not take action until the Republican political machine took control of local politics in the 1890s, and even then only because the machine could profit from new sanitary measures.[62] Pittsburgh's machine took advantage of the garbage problem. Although Pittsburgh was ruled by a very different corrupt regime, as in Charleston, garbage was yet another way for elites to enrich themselves.

In 1895 the Pennsylvania legislature created bureaus of health in cities of the second class (those with population between 100,000 and 1 million) to address conditions that would imperil public health, including garbage, rubbish, dead animals, privy waste, and tainted food.[63] The director of Pittsburgh's department of public safety was enlisted to draw up a general sanitary code for the state, and state health authorities discussed issues from compulsory vaccinations to hygiene, typhoid, rural public health, and the proper use of disinfectants.[64] The state authorized the director of the department of public safety (within which the bureau of health was located) "to enter into a contract or contracts with such parties as may be found necessary, for the removal of all offal, garbage and swill from private premises and for the disposal of the same."[65] Although the legislation seemed beneficial to the public welfare in cities generally, by specifying cities of the second class, the new law pertained only to Pittsburgh and neighboring Allegheny (which Pittsburgh would annex in 1907). In practice, the legislation legally empowered the machine-controlled department of public works to distribute public money to machine-affiliated contractors. In other words, it was written by the state legislature to benefit the Magee–Flinn regime.

The city council wasted no time acting on this authority, passing an ordinance to provide for the "collection, removal, and disposal of garbage, offal, dead animals, and condemned meat in the city of Pittsburgh" by contract and at public expense. The director of the department of public safety was authorized to enter into a contract or contracts with private parties.[66] The first four-year contract was awarded to Charles E. Flinn—machine boss William Flinn's brother—and was even renewed in 1899 for the sum of $93,890 per year.[67]

It is no coincidence that the City of Pittsburgh addressed a very real garbage problem by contracting out to a machine-owned (indeed family-owned) business. Yet it was not a total boondoggle. Unlike Charleston, which relied on antiquated technolgies, Pittsburgh was innovative. Charles Flinn ran the American Reduction Company, which, in addition to collecting trash, was also tasked with disposing of it. Flinn's company invested in a state-of-the-art reduction facility, an innovative technology for extracting reusable products from garbage.[68] The plant cost upwards of $100,000 at the time.[69]

Reduction was an elaborate large-scale process. First, garbage was put into large steel tanks called "digesters" and "cooked" for six to eight hours. The cooked garbage was then pressed in hydraulic presses. The water and grease mixture that was extracted was sent through settling tanks, which removed the grease while the water flowed into a river or sewer. The grease was mixed with naphtha or some other solvent, then separated again through vaporization. Meanwhile, the pressed garbage was sent to a dryer and then bagged for shipment. This process rendered garbage sanitary and its by-products suitable for reuse. It did smell. Reduction plants gave off "a strong caramel odor, approximately described as the odor of sweetish burnt coffee, and which is very objectionable to many nostrils."[70]

Reduction was cutting-edge technology for its time, and an innovation that sought to make money off the by-products of waste. It was one of the processes endorsed as "the promising methods of future garbage disposal" by the American Public Health Association in 1889.[71] In choosing reduction as its method of disposal, Pittsburgh was among the innovators of its time. The digesters and presses and dryers were a feat of engineering that required both expert knowledge and economic investment. Engineers were eager to build and oversee these plants, and they chafed at cities' unwillingness or inability to invest the funding. In 1902 the *Municipal Year Book*, a survey of public works around the country overseen by an editor of *Engineering News*, noted "the stubbornness with which most American communities cling to primitive and unsanitary methods of garbage disposal," bemoaning the fact that of the 1,524 cities and towns surveyed, only ninety-seven used the modern methods of reduction or cremation.[72] City governments generally lacked the capacity to build and operate reduction plants. Using the power of contracting out, they relied on the capacity of private resources. Unlike Charleston's antiquated, backwards-looking administration, which relied on exploitative labor practices, Pittsburgh's machine relied on business resources and saw innovation as profitable.

Early reports indicated the machine's reduction plant was a success. William Flinn bragged that unlike plants in Philadephia and New York, which released water into rivers and allowed garbage gas to escape, the Pittsburgh plant had found a way to "destroy the gas."[73] Eighty to one hundred tons of garbage were hauled to the furnace daily, and 120 barrels of residuum were produced in exchange, with "no offensive odors at any time, except when the fresh garbage is brought in."[74] Flinn was pleased with the quality of the grease that the reduction plant produced, without excess water or impurities. The first profitable car of grease sent to Proctor & Gamble in Cincinnati net a profit of $921.[75] Flinn wrote to leaders in Rochester and Washington, DC, offering partnerships to build and operate similar reduction facilities in their towns.[76]

Despite the initial promise of the 1895 ordinance, garbage collection flagged over time, and it was evident that the plant would not be the profitmaker it promised to be. The quality of grease was inconsistent.[77] Flinn admitted that there were few workers employed at the plant, and those there were did not work full-time.[78] Complaints mounted about Charles Flinn's irregular collection practices.[79] Ashes and dirt—which were not collected with the trash—were stored in unregulated barrels, and their removal depended on the householder. Filth notices, which were orders to procure garbage pails, were often given out but never followed up on. Rather than properly designed, standardized garbage receptacles, Pittsburgh had "unfit and impossible-to-keep-clean contrivances."[80]

The Civic Club of Allegheny County kept track of the complaints. Formed in 1895 the Civic Club was committed to sanitizing both the city's streets and its corrupt government.[81] The first activities of the club focused on achieving pure water and garbage disposal.[82] The Civic Club drew attention to the poor administration of garbage collection, keeping lists of complaints about alleys and sanitary conditions, particularly on the working-class South Side.[83] Public health officials reported that the dump did not have enough watchmen, and those employed were not uniformed. Carts were not covered with canvas as required.The weighing of cart loads was not supervised by a weighmaster, allowing garbage collectors to overcharge the city.[84] Even though garbage collection was uneven at best, and even though the American Reduction Company's failures were well documented, it still continued to receive the garbage collection contract.

Effect of Corruption in Integrated Cities

In both Charleston and Pittsburgh, the corrupt, informal governing regime was able to rule the city and benefit itself, often through legal (albeit unethical) means. Although Charleston had one of the earliest trash collection programs in the country, it did little by way of innovation in the years following its inception.

In 1887 Charleston's board of health reported employing a satisfactory scavenger service for routine city services, with thirty carts in operation to haul away garbage, clear drains, and fill in land, as the city had done earlier in the century. The scavenger service was also useful in emergencies, helping to clear debris after storms, cyclones, and earthquakes. While the board was pleased with city ownership of scavenger services, it did suggest that the city use an incinerator to dispose of garbage. Charleston's standard method was to dump it in the salt marshes outside areas of human habitation, where the tide would wash it away. The board of health regularly urged a more modern method of disposal upon the city.[85]

Charleston used convict labor for public works, particularly for the most physically demanding tasks. To ensure a robust supply of workers, police cracked down on petty offenses. Rather than dole out public jobs to new immigrants as political machines did, Charleston's elite—who were under no democratic pressure—had been content to assign administrative offices to families in the inner circle and to farm out work to Black laborers, shifting to convicts once slavery ended as a source of labor. The system played a prominent role in the economic expansion of the New South—from railroads to coal mines to roads. Defined by the exploitation of overwhelmingly Black labor and "steeped in brutality," it became the new form of forced labor after emancipation.[86]

As in other southern cities, leaders in Charleston worked to reinstate a prewar racial order. In 1885 South Carolina passed a law allowing convict labor and empowered municipal authorities to impose punishment on laborers within their respective jurisdictions. Charleston did not use convict labor right away. Under Tillmanite mayor John F. Ficken, who garnered the support of Democrats united in their desire to weaken the Ring's control over city politics, the "Convict Force" was inaugurated in the city of Charleston in 1892. In the first year the city used convict labor to level sidewalks, clean and grade streets, and clear ditches and trim bushes alongside streets. In 1893 the South Carolina Supreme Court in the case of *State v. Williams* ruled that a Court of Trial justice lacked the authority to impose a sentence of hard labor. Those convicted in the Court of General Sessions were usually guilty of more serious offenses and were sent to the state penitentiary. In 1894 the chain gang was abolished. The street department was "deprived of this class of labor, and the work of grading and improving the extreme upper portions of the city [was] curtailed."[87] This decision was overturned in 1895, when the new state constitution gave courts the power to sentence convicts to hard labor.[88] In 1899 the General Assembly passed an act that authorized the utilization of county chain gangs for the kinds of work "calculated to promote or conserve public health." The city used an average of forty convicts a day; in 1899 786 laborers were Black, eighty white.[89]

In another amendment to the state statutes enacted by the General Assembly in 1901, the county chain gang was placed under the Sanitary and Drainage Commission of the County of Charleston, and convicts were tasked with draining and cleaning, reshaping, and re-leveling twenty thousand feet of county ditches. The city convict force graded, leveled, and drained, filled up all low places, cleaned roads and shell walks, cut down dead trees, burned trash, opened ground for the laying of pipes, and dug ditches, among other tasks.[90] The implementation of the vagrancy laws provided a steady stream of labor, overwhelmingly Black.[91] The 1903 "Report of the Commissioners for the Management, Care and Custody of Convicts" detailed receiving 754 men to work on the chain gang. Of these, 660 were Black (88 percent), while ninety-four (13 percent) were white.[92]

When Broad Street candidate R. Goodwyn Rhett assumed office as mayor in 1903, he made major changes to garbage policy. He established a board of public works and placed garbage collection under it. He purchased new city carts and new mules. And he required drivers to be uniformed and increased their wages. Unpopular with residents, Rhett reduced garbage collection to three or four days a week and, in 1907, limited the types of refuse the city would be responsible for removing. Chain gangs, however, were still used for hard labor, and vagrancy laws, which were implemented with "vigor," were credited with decreasing the number of "idlers and country negroes on the streets."[93]

The Broad Street Ring was dealt a blow when Mayor John P. Grace came into office, serving two nonconsecutive terms, 1911–1915 and 1919–1923. One of Grace's first acts was to dismantle the department of public works created by Rhett, claiming it was "too political."[94] There was significant municipal improvement under his administration. Noting that the practice of disposal by dumping in salt marshes would become untenable as the city grew closer to the marshes, he considered the use of a high-temperature incinerator, the Destructor, but its cost was prohibitive.[95] The administration of Mayor Tristram Hyde—who held office between Grace's two terms—went ahead and made a contract to erect and operate the Destructor.[96] Nevertheless, the Destructor needed frequent repairs and was not able to operate for the nine hours a day needed to incinerate all the trash. The street department blamed this poor performance on the garbage collectors' failure to bring enough garbage to heat the Destructor sufficiently.[97] The Hyde administration may have made a pretense of innovation, but it failed to follow through with adequate operation and maintenance. Grace complained upon his return to office, "I found to my dismay every department functioning on the lowest possible basis." He pronounced the situation "a municipal disgrace." As for the street department, "we found things as we left them," reported the mayor, citing four years of neglect and the need to provide new equipment.[98] Grace was not able to remedy the situation. Under Grace's regime, the Destructor was unusable.[99] Grace was derided as a boss even while he had a reputation as a progressive. In our own research and that of Jessica Trounstine, we do not make too fine a distinction between the two.[100] Whether progressive reformers or political machine, both types of government sought to maintain their power, which was achieved at the time by claiming (or actually following through with) public improvements.

While other cities were finding new ways to address the garbage problem, Charleston's elite continued to do what they had done for decades, employing friends and relatives to oversee departments and using inexpensive, racially exploitative labor for city services. When Charleston did finally try new garbage disposal techniques, in the early twentieth century, the Destructor failed miserably. Unlike Pittsburgh, where the Magee–Flinn machine had the capacity to provide services to the city, Charleston could not rely on its informal regime to build its capacity.

Also, unlike Charleston, Pittsburgh was able to parlay the machine's infrastructure investment into a beneficial force for the city. After Christopher Magee's death in 1901, the machine fell apart. Flinn ousted E. M. Bigelow from the public works office in a political dispute. Bigelow rallied disempowered reformers, Democrats, and others to take back the government.[101] As the new boss in Pittsburgh, Bigelow captured the old Magee-Flinn organization and merged it with the reformist Citizens Party.[102] Reformers saw their opportunity.

In his quest for change, Mayor George Guthrie ran into the power of the old machine. When the administration prosecuted six public officials for tax fraud, the defendants were acquitted. When the administration obtained the resignation of the director of public safety, he was given a new position as mercantile appraiser. The *Pittsburgh Press* announced, "Guns of the Flinn and [George T.] Oliver factions in the Republican party were unmasked today—War has started."[103]

Pittsburghers fought back against the corruption. Seventeen civic organizations came together to draw up a new charter in 1910. The city's residents supported it, and fifteen thousand petitioned the legislature to pass it unchanged. In place of cumbersome city councils with members representing individual wards, a system ripe for corruption, by 1911 Pittsburgh adopted a nine-man council with citywide districts replacing wards.[104] In the 1910 election voters approved a $100,000 bond to acquire land for erecting and equipping a municipal incinerating or refuse disposal plant.[105] Nevertheless, in 1912 it awarded the garbage collection contract for the city of Pittsburgh once again to the American Reduction Company for $185,700. Although annexed by now, Allegheny had its own contract, awarded to the Allegheny Garbage Company.[106] Even though Pittsburgh was controlled by reform regimes, problems remained with garbage collection. City officials blamed residents for failing to comply with instructions.[107]

More significantly, the two reduction companies continued to receive the garbage contract thanks to the triumph of the Democratic Party in the 1930s. The city and civic groups began looking into city control of garbage collection and erection of a city-owned incinerator in

1933, appointing to the waste disposal committee a mixture of academics, medical experts, businessmen, and the sanitary engineer from the American Reduction Company.[108] Until that incinerator was built, the city continued to rely on the American Reduction Company. Although the company overcharged the city, it enjoyed a virtual monopoly over the city's contract until the end of the decade.[109] By 1940 the city had an incinerator and was ready to dispose of its garbage. But it did not have enough trucks to collect garbage, so the American Reduction Company "rescued" the city by renting out its trucks.[110] By the time the city was running its own trucks, truck drivers were "lined up for blocks" at the incinerator on overtime pay because of delays in unloading, as the city had purchased the wrong kind of truck and had to unload "practically by hand."[111]

Capacity

Corruption provided the incentive to establish municipal collection in Charleston and Pittsburgh, with Charleston using slavery as a resource in the early nineteenth century and Pittsburgh using the Magee–Flinn machine's business connections at the end of the century. In both cities, the form of corruption was integrated into municpal government, with Charleston's aristocratic practices woven into city governing and services and Pittsburgh's machine dominating city government and well established in state politics, too. That integration produced actual collection. Charleston's elite kept their circle tight by not relying on outside actors. Pittbsurgh's machine had an incentive to innovate with a novel reduction plant in order to claim the city contract. The experiences of Charleston and Pittsburgh suggest that corruption that is integrated with government can be a resource for providing capacity.

Even so, the stories of Charleston and Pittsburgh are rife with undemocratic elements. Charleston may have been an early adopter, but its development of city services on the basis of slavery and, later, on racial hierarchy incorporated inequality into this capacity. Administratively, Charleston's long history of garbage collection showed signs of strain as the rest of the country caught up in municipal collection. While cities across the country innovated, Charleston's collection system remained relatively stagnant. The city's ruling families benefited from employment in city offices and through renting out their slaves to do much of the city's work. The aristocratic leadership, however, was content to rest on its laurels. Rather than continue to innovate or to develop new models of trash collection and disposal after slavery, the city continued with a modified version of what it had done for decades. By contrast, Pittsburgh's system of collection was suddenly innovative, emerging from decades of sanitation efforts based on nuisance laws with investment in a startling new method of collection and disposal by reduction when the Magee–Flinn machine came to power. This innovation was remarkably sustainable, outlasting the machine itself. The American Reduction Company continued to receive the garbage collection contract long after regime change and political reform, and it provided capacity until the Democratic Party could do the job on its own.[112] The Pittsburgh machine was not mass-based, and it dimissed reformers, signaling that it was unresponsive to the people. Nevetheless, in both Charleston and Pittsburgh, democracy and progressive governments were not required to generate the political will to do something about the garbage problem and to create comparatively successful programs.

CHAPTER 5

Solving the Garbage Can Problem

Race, Gender Hierarchy, and Compliance

Corrupt city governments managed to get trash collected, each with its own reasons for taking on garbage collection and sanitation practices. While it could decide whether to use city collection, or contract out, or rely on outside services, a moment inevitably arrived when a city needed householders to put out their garbage cans. Garbage had to be put out at the right time, in the right spot, in the right container, and sorted in accordance with that city's disposal method. Implementation of garbage collection did not rely just on the work of city actors or contractors or entrepreneurs; it relied on the cooperation of private residents. These residents had not been welcomed into regimes, and in fact the more engaged of them had been actively pushed out. Corruption was not necessarily a problem for establishing garbage collection, but now cities had a garbage can problem, and the solution lay in residents who may not have been invested in solving it.

Facing this plight, governments turned to the technical experts and civic associations that had been so eager to establish municipal garbage collection in the first place. Those groups gladly participated because in doing so they could achieve their goals and benefit as organizations. Engineers willingly responded to cities' requirements for technical knowledge and infrastructure, taking advantage of opportunities for the profession that came with the rise of city planning and city services. The benefits were material and immediate, personally and professionally, with the awarding of contracts or the securing of positions in departments of public works. But technical expertise was not enough to solve the implementation problem. White women's civic organizations that had been pushed aside by corrupt regimes likewise found an entree into garbage collection when cities faced challenges in implementing municipal services. In modeling behavior, they could parlay their race and class privilege into distinguishing themselves as model citizens. Gender, then, intersected with racial privilege, emerging as another informal resource of city governments in carrying out their programs.

Engineers' Technical Expertise

Engineers could benefit from garbage collection practices, corrupt or not, through contracts and through placement in departments of public works or public utilities. There were plenty of opportunities to profit from the city planning and city services that emerged from municipal development. Engineering professionals realized that "municipal officers of American cities control and supervise expenditures exceeding those of any one private industry."[1] Municipal political development was a gold mine for engineers.

As garbage collection programs gained traction across the country, the engineering profession was changing. Traditionally, an engineer would be in charge of all aspects of a project; as industries consolidated and projects became larger in scale, engineers began to take more of a managerial role, hiring more assistants and overseeing technical workers.[2] The *Engineering News* compiled statistics on the 1,524 cities of over three thousand inhabitants in 1902. While there were officials overseeing municipal projects in these cities, the editor lamented "a lack of uniformity of method and of standards for similar work under like conditions that make the annual reports of little value for purposes of comparison."[3] Engineers could fill that gap and standardize resources across the country.

The American Society of Municipal Improvements brought together city officials and engineers in street pavement, street lighting, water supplies, sanitation, garbage disposal, public utilities, street cleaning, sewage disposal, and other municipal projects. The organization allowed for networking and for learning about innovations and opportunities in other cities. Members could share the "best methods of selecting paving brick . . . the most satisfactory and economical system of cleaning streets, of collecting garbage and ashes . . . the best manner of regulating the too promiscuous opening of street pavements," and other projects.[4]

Advertisements also disseminated information. When a reader of *Municipal Engineering* submitted a letter in 1901 requesting the names of garbage crematory manufacturers, the editors directed him to the issue's advertisements, which served as an index of the latest technologies.[5] A "Recent Inventions" section in the journal shared news of innovations such as a garbage can with tapered sides so that it could easily nest within another.[6] The journal also posted calls for proposals and notices of contracts to be awarded for various municipal projects, including garbage disposal. A 1901 issue, for example, announced that a garbage crematory would be built in Springfield, Ohio. Another crematory site was authorized in Spokane. The question of building a garbage crematory was being discussed in Findlay, Ohio.[7] With such news, engineers could keep apprised of opportunities down the road and be prepared to submit a proposal.

While the need for garbage collection had been promoted early on by sanitarians, it was engineers who seized opportunities to advance their technology and their profession. According to an article in *Municipal Engineering:* "The recovery of the values in the refuse and garbage is the point of greatest interest in modern methods, developed mainly since the subject has been made once for consideration by engineers rather than by physicians, and the possibility has been demonstrated that collection and disposal can be made without nuisance and in a perfectly sanitary manner."[8] Nevertheless, garbage collection continued to follow the sanitary model. Engineers certainly were pursuing innovations that could extract by-products from waste which could be sold at a profit, as with reduction. The guiding framework behind garbage collection, however, continued to be sanitation, with waste considered a hazard.[9]

The Garbage Can Problem

Engineers had access and opportunity. They were able to implement their technological advances, yet all this work could be stymied if people did not put their garbage cans out or did not do so in keeping with the methods required by the technology. Referencing the travails of New Orleans in 1895, *Engineering News* pointed out, "One of the greatest

obstacles to the sanitary disposal of garbage is the indiscriminate collection of ashes, tin cans, glass bottles, old clothing and other refuse
which finds its way into the garbage can." Despite ordinances requiring that different types of refuse not be mixed, and despite garbage
contractors refusing to take it, "many householders adhere to the old
custom of dumping everything into the garbage can, and complaints
are loud if it is not emptied."[10] If the disposal method could handle
only certain materials, then these experts counted on householders to
sort their own garbage and put out only those materials that could be
processed. Corrupt regimes decided to collect garbage, and engineers
were willing to land the contracts or make their inventions available to
cities. Garbage trucks and drivers and collectors were sent out to collect
the garbage. But for them to do their part, residents had to put their
garbage cans out in the way the city required, and householders were
slow to do so. Thus cities faced a garbage can problem.

Requiring residents to put out their garbage cans was to ask them
to practice new habits, to comply with new rules, and to change their
routines in their homes. This was being requested of them by a regime
that was likely corrupt and was not terribly concerned about fostering meaningful public support. If regimes were not concerned with
democratic legitimacy, and they were not wooing the public, then why
would people heed their very specific rules about where to put their
trash and when? It was a lot to ask of residents who, before municipal
garbage collection, had gotten rid of their household waste by burying, burning, or dumping it whenever and however they wanted. At
the household level, municipal garbage collection requirements may
have seemed more of a burden than the garbage problem itself. Consider the imperatives placed upon residents when Louisville reformed
its garbage collection practice in 1918 (figure 5.1). Householders had
to obtain their own garbage receptacles, which were to be "metal,
water-tight, with close-fitting covers and with handles sufficient for
convenient emptying." These cans needed to be of regulation size,
"large enough to hold four days' accumulation" for a family of six.
(The number 9 Witt or King cans were recommended.) As for contents, the garbage can was to contain only animal and vegetable waste
from the kitchen—not cans, glassware, crockery, paper, or cuspidor
waste. All receptacles were to be placed in the alley or at the curb to
be ready for collection. Once the garbage was collected, the can was to
be brought back inside to the kitchen immediately and cleaned with
boiling water.[11]

GARBAGE
Do's and Don't's
THINGS TO DO

DO—Drain garbage of surplus water, and WRAP it in heavy paper or several sheets of newspaper before placing it in a container. This keeps your container more sanitary—keeps garbage from freezing to containers—prevents garbage acids from destroying container and prevents practically all odors.

DO—Use separate containers for combustible (burnable) waste and non-combustible (unburnable) waste.

DO—Place the proper container for collection on EVERY collection day.

DO—Use water-tight containers with tight-fitting covers.

DO—Keep containers always covered.

THINGS NOT TO DO

DON'T—Put any liquids in containers.

DON'T—Permit waste to overflow containers.

DON'T—Sweep, rake, or place refuse onto sidewalks, lawns, or streets.   Put it in proper containers.

FIGURE 5.1.   Garbage rules
Source: Garbage Clippings, Garbage Disposal and Incinerators, bk. 5, Louisville Free Public Library.

Even when can materials were specified, such as metal, earthenware, or galvanized iron, most cities saw garbage put out in "wooden barrels, firkins, tubs, pails, and boxes" without adequate covers. And then there was the problem of pickup location. From a public health standpoint, it was best to place the garbage receptacle as far away from the house

304   MUNICIPAL

FIGURE 5.2. Evidence of residents' noncompliance with garbage ordinances
Source: *Municipal Engineering* 52, no. 6 (June 1917): 304.

as possible, but few cities required that. After all, if householders were responsible for carrying out the garbage pail, then they had little incentive to travel far with it. Laggard attitudes toward garbage cans explain why, in his comprehensive review of garbage collection and disposal published in 1901, Charles Chapin identified garbage receptacles as "one of the commonest [sic] forms of nuisance."[12]

Municipal programs, with their engineering marvels, administrative commitments, drivers, collectors, horses, and carts, were not going to improve upon the past if people failed to put their garbage out as required (figure 5.2). When cities passed their new garbage ordinances, they were replacing more primitive disposal methods and lax customs. Pittsburgh could rely on more professional workers when it invested in its garbage contract specifying up-to-date disposal methods in 1895, but the use of cans was slow to catch on. In 1913 the New York Bureau of Municipal Research's study of Pittsburgh indicated that "suitable garbage cans with covers are frequently lacking in the houses." Despite having inspectors on hand, people did not understand the instructions, or did not adhere to them, there was no uniform use of garbage pails, and there was inadequate prosecution for lack of compliance.[13] The formal garbage collection ordinance and accompanying administration did little to modernize compliance by households. Householders across our seven cities greeted new regulations with outright refusal, resistance, or foot-dragging.

Cities did have some enforcement powers to deal with noncompliance. Traditional nuisance laws provided them with the authority to punish (primarily through fines) those whose waste was a hazard to others. Nineteenth century boards of health used this authority to issue fines for sanitary violations. The most advanced cities set up a team of health inspectors to locate and punish offenders. New York City instituted house-to-house inspection, conducting 42,909 inspections in 1896 and abating 38,858 nuisances.[14] When conducting inspections, health departments recorded their work in a street index, using a form listing the address, date of inspection, owner, condition, date of complaint, and action taken. Violators needed to be given due notice. Health inspections recognized the fact that effective garbage collection does not start in the street with the city collectors but begins inside the home. Residents were required to separate their trash, place it in the right cans, and put it in the designated spot on the street on the appointed day and time. To make that happen, residents needed to reorder their own consumption, use, and production patterns in the kitchen and home. Hence, city garbage collection required certain kinds of behavior in people's homes, out of sight of the state but not out of its reach.

Modern statecraft often requires governments to reshape citizens' behavior. Modern governments that erect new policies require a citizenry that is prepared to be regulated or acted upon.[15] The citizens and city space may need to be rendered legible, standardized, so that they are capable of being counted, extracted, or regulated. Other kinds of regimes could have made use of this far reach of the state, but surveillance via health inspection required resources that many cities did not have or were not willing to expend.

Heavy-handed use of power, if enacted upon a citizenry not connected to the government, presented a legitimacy problem. Cities that rested on corruption could not be guaranteed of winning the hearts and minds of its citizens. Residents who preferred to retain the old means of garbage disposal were poised to resist intrusion into their habits by city government. When New Orleans contracted with the out-of-state Southern Chemical and Fertilizer Company to collect garbage and dispose of it in a state-of-the-art garbage reduction plant in 1894, New Orleans residents refused to cooperate. They were accustomed to city carts collecting their waste, a system that was ineffective but famil-iar. Residents were not pleased to learn that Southern Chemical would be in charge of removing waste. They complained that the contract was unfair and that the rules they would have to follow were too onerous. In particular, they bristled at the requirement to purchase regulation garbage boxes and the threat that they would be fined if they failed to do so. The *Daily Picayune* noted, "The garbage box ordinance is being dis-cussed everywhere, especially among the housewives, on whom the chief burden of responsibility is attached." Housewives found the required garbage boxes to be "an unmeasurable source of expense, hardship and nuisance," declared the *Picayune,* which concluded, "The garbage-box requirement is justly becoming a serious question."[16] Despite the harsh penalties for noncompliance, residents simply did not rush out to pur-chase the requisite garbage boxes.[17] The garbage box issue was so ubiq-uitous that it earned its own ode, which reached *Picayune* readers on March 18, 1894:

To the Garbage Box

Oh, garbage box! Oh garbage box!
So beautiful and neat,
It is a sight to touch the heart
To see thee, patient as thou art,
Standing all day upon the street,
Waiting for the garbage cart
To cull thy treasures sweet.[18]

New Orleans's garbage contract was blatantly corrupt (and eventu-ally a political liability for Mayor Fitzpatrick), and complaints about the garbage box may have been a pretext in a partisan fight.[19] Never-theless, they had a legitimate resonance: the city couldn't tell people what kind of trash can to use, and if the city paid someone to pick up garbage, he should pick up all of it. When the cost of the contract was at issue in a grand jury investigation, the jury's "main objections were the same which could be heard on the streets of the city—the definition of 'garbage' which the contractor would pick up was too limited. He refused to cart off old shoes, bottles, broken glass, or tin cans. A special type of receptacle had to be used for garbage, or it would be left to rot on the sidewalk."[20] Faced with massive public resistance, the contract ended, as did Mayor Fitzpatrick's regime. Acts of resistance and commonsense complaints served as a persuasive form of popular protest against this new exercise of state power.

Infrastructural Power

New Orleans's residents recognized that the city needed their cooperation, and they knew that they could flout the rules with foot-dragging and resistance. Inducing compliance to garbage can rules required cities to exercise power—to get people to do what they otherwise would not. Cities were in the delicate position of ordering people to change the way they had always done things and, moreover, to change the way they had done things in their homes. It was too intimate a space for the state to enter with such an authoritative approach. Cities were better off taking a more diffuse approach and operating instead with softer forms of power, such as ideological power, which would shape a community's shared norms and values.[21] If compliance with garbage collection could be circulated through norms in civil society, then people would internalize government imperatives as habits in keeping with their peers. Governments could then rely on infrastructural power—using economic and social relations on behalf of the state—to enter into the realms of civil society and the home. They would obey without the city government looking despotic.[22] The new practices would look like a habit rather than a command. Habits at the curb started in the home, and women reformers were well suited to move between these spaces. Mary McDowell—the activist instrumental in improving Chicago's garbage collection—pointed out in a speech she gave in Charleston that garbage posed a housekeeping question. A good housekeeper was known by the kempt condition of her back door and kitchen and alley. So too should cities be judged by the tidiness of their proverbial backyards.[23]

McDowell's sentiments reflect both the ability of women's groups to make a public contribution and their distinction as sanitary citizens. Women's civic groups followed the "municipal housekeeping" doctrine, which implied that a woman kept her home clean, and kept right on cleaning until she was sweeping the street, extending her virtuous efforts into public spaces. It was a way for women to enter the public sphere in the guise of domesticity.[24] The garbage can problem allowed municipal housekeeping to work in the other direction. Women could bring proper garbage can practices into the home, extending the public into the private sphere.

This process was facilitated by companies that devised new sanitary items for household use. Readers of the May 1917 issue of *Good Housekeeping* were offered the Wayne Garbage Bag and Stand, a sturdy metal stand for holding garbage bags. The bag was an innovation that the housekeeper could just throw away with the garbage, rather than go through the "nuisance" of having to wrap the trash. "How Convenient and Sanitary!" indeed. The Wayne Garbage Bag was touted as "A Necessity in Clean Kitchens." As for the garbage can that the

FIGURE 5.3. Innovations included the Witt's Can & Pail
Source: *Good Housekeeping*, June 1918, 127.

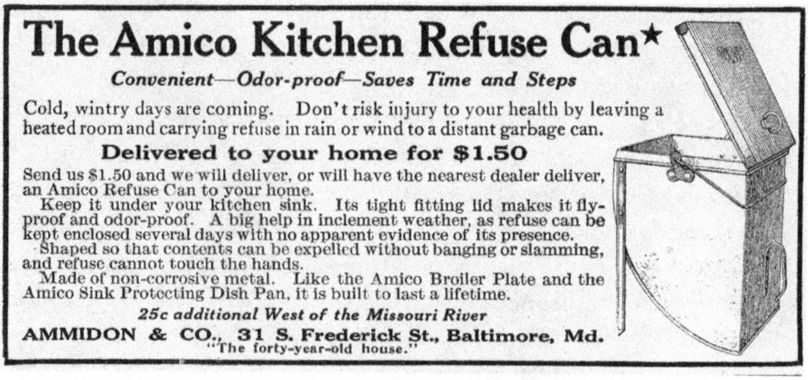

The Amico Kitchen Refuse Can★

Convenient—Odor-proof—Saves Time and Steps

Cold, wintry days are coming.   Don't risk injury to your health by leaving a heated room and carrying refuse in rain or wind to a distant garbage can.

Delivered to your home for $1.50

Send us $1.50 and we will deliver, or will have the nearest dealer deliver, an Amico Refuse Can to your home.

Keep it under your kitchen sink.   Its tight fitting lid makes it fly-proof and odor-proof.   A big help in inclement weather, as refuse can be kept enclosed several days with no apparent evidence of its presence.

Shaped so that contents can be expelled without banging or slamming, and refuse cannot touch the hands.

Made of non-corrosive metal.   Like the Amico Broiler Plate and the Amico Sink Protecting Dish Pan, it is built to last a lifetime.

25c additional West of the Missouri River

AMMIDON & CO., 31 S. Frederick St., Baltimore, Md.
"The forty-year-old house."

FIGURE 5.4. Another innovation: the Amico Kitchen Refuse Can
Source: *Good Housekeeping*, October 1916, 194.

housewife needed to set out at the curb, she had choices and aids here, too. The Witt Garbage Can (recommended by New Orleans officials) promised a tight-fitting lid on top of a can that was fire-, odor-, dog-, leak-, and scatter-proof (figure 5.3). Furthermore, its heavy galvanized steel and deep corrugations made it twenty-nine times stronger than plain steel. The Amico Refuse Can was a metal container that could be kept under the kitchen sink until the garbage was ready to go out, saving the housekeeper multiple trips in inclement weather (figure 5.4). While that garbage was sitting indoors in one of these various receptacles, another product, Lysol, allowed the housekeeper to prevent disease "the same way big hospitals do it." Among its many uses was sanitation of the garbage can, a breeding ground for germs in a place where children play.[25]

One way to solve the "garbage problem" was to make each individual household rather than the city responsible for collection and disposal. *American Kitchen Magazine* (advertising itself as "A Domestic Science Monthly") featured an article about a world where technological improvements made swill pails and carts unnecessary. It recounted a boy's dream of pneumatic tubes under houses, whisking garbage out of sight, adding, "There seems at present to be no escape from garbage men and carts, but at least we can make matters more satisfactory by fully understanding what we have the right to demand."[26] The same issue featured a home-based garbage dryer. Given that incinerators could not always operate at capacity, and dumps were breeding grounds for disease, "it now appears as if the opportunity to solve this question would be put into the hands of women." The dryer could be attached to the kitchen stove or chimney, making safe a practice that many astute housekeepers already followed in their own homes—drying and burning vegetable parings to reduce household refuse. While "the house-cleaning proclivities of women have long been the jest of the newspapers," the article acknowledged that each woman, performing this work one by one on her kitchen stove, would serve the public good, save cities money in garbage collection, and contribute to the sanitary well-being of all: "If every householder sees to it that the household refuse is properly sorted,—the ashes, broken dishes, tin cans, etc., in one receptacle, and the perishable portions in another,—the whole matter would be simplified. Moreover, if all could be reduced to ashes, there would be much less bulk to be removed, and no harm would come if it were not collected daily."[27]

By keeping dogs and vermin from dispersing the garbage, the Majestic Garbage Receiver provided the housekeeper the chance to protect her family from flies breeding. But really, this was just an instance of modern materials—a watertight metal container—evoking a primitive form of garbage disposal: burial. The Incinerite also updated a primitive method by providing a receptacle for indoor incineration: "In homes where health, cleanliness and convenience are first considerations, the trouble and exposure of carrying garbage to an unsanitary, germ-breeding garbage can is displaced by . . . The Incinerite." Designed to replace the garbage can in the home, the Incinerite was touted as "The Invisible Garbage Man," allowing the housekeeper to avoid the garbage collector altogether. In this formulation, women took charge of their own garbage, and they had the resources to accomplish that in their own homes, on their own. They were able to bypass collection and go immediately to the work of disposal. And they could be trusted to do this work responsibly. Another device, the Sani-Can, was a covered enamel can with removable pail to be kept in the kitchen. Pressing the foot pedal opened the lid, allowing the housewife to empty waste "without stooping." Releasing the pedal not only closed the lid but also sprinkled disinfectant over the contents. This "ideal Christmas gift" had the properties of both sanitation and convenience (figure 5.5).[28]

Manufacturers were opportunistically playing into a newfound notion of sanitation, but they were also returning the public garbage problem back into the home. If compliance rested on individual behavior, then the ideal housewife could serve as the model of a good citizen with a clean kitchen and appropriate garbage practices. The

FIGURE 5.5.   Sanitary household conveniences
Source: *Good Housekeeping*, December 1916, 158.

woman-as-consumer could contribute to solving the public garbage can problem in the private sphere.

Women's Civic Organizations

Local governments had largely kept women reformers at arm's length, but they could harness that enthusiasm in their exercise of infrastructural power. This stealth form of power had ready operatives—civic-minded, organized women who had initially been pushed out of the decisions surrounding garbage collection. Women's civic organizations were ready to serve. They had been engaged in the subject of garbage collection from the outset, with a keen interest in creating garbage collection programs in the 1890s but rarely included in those programs. Such women were not a monolithic group. Opportunities to convert their civic projects into government access depended on their own position in the city and its regime. In cities across the country, civic clubs were prominent in the implementation of municipal garbage programs.

Charleston Civic Club

The Civic Club of Charleston was unable to find a place for itself in formal garbage collection, so it poured its sanitation efforts into supplementing programs, providing services not met by the city. The Civic Club organized to make a difference, establishing both health and betterment committees that worked to clean up the city, but it was unable to get involved with Charleston's garbage collection. It focused its sanitation efforts into purchasing and maintaining public trash receptacles, organizing a Clean-Up Day, and photographing "unsanitary places." The Civic Club claimed credit for the 1904 ordinance requiring the use of covered trash cans; the establishment of a Junior Civic League that purchased public trash receptacles in 1909; and a "fly" campaign, contests for white and Black children, separately, in which money prizes were awarded for the killing of flies.[29] It took additional measures on behalf of city officials, placing seventy-five trash cans around town. The Junior Civic League also worked with the street department to obtain funds to make repairs to cans damaged by wear and tear.[30] This work was organized by race. The Charleston Civic Club, an organization for white women, had had the idea of working with Black women on schools and housewives' issues, but once decided, "this feature had been turned over to the colored Y.W.C.A."[31]

The Civic Club had the power to raise awareness and direct officials to areas of need. As civic clubs did in other cities, the Junior Civic League circulated instructions to householders, this time in a newspaper announcement headed "The Tin Cans of the Junior Civic League," calling upon citizens to tackle the wastepaper that was flying around the city. They asked the "good housekeepers" of Charleston to help sanitation workers by burning their paper trash rather than disposing of it, so the paper would not fly about as the carts lurched through the streets.[32]

Charleston's Civic Club lacked access to city offices, however. When club officers wrote to officials such as the chief of police and the head of the street department, they were received cordially but not welcomed to get involved in city business. The reason may lie in the street department's lack of interest in doing the job effectively. Charleston had ordinances on the books that had been there for years, but it was not able to implement the laws in place, given the lack of coordination between the street and health departments and a disinclination to take up the Civic Club's offers of assistance.

In 1920 the street department reported a garbage can problem, noting that the requirement to use watertight metal receptacles, with half-flour-barrel capacity and a tight-fitting lid, set out at appointed hours, was honored mainly in the breach. Liquids and scrap paper were being placed in garbage bins, as were lawn cuttings, tree trimmings, and rubbish. Residents used all sorts of receptacles—fifty-gallon barrels that required two or three men to lift, uncovered coal scuttles, soap boxes, old trunk tops—left out uncovered, and frequently after the garbage cart had passed. The street department lamented, "From all of this it is evident that the refuse collection service, than which probably no municipal activity touches more directly the citizen body, does not receive from our citizens that general co-operation to which its importance should entitle it."[33] Despite needing help, and the presence of women's groups eager to fill in, Charleston relegated women's participation to providing public trash cans and did not incorporate women into the municipal collection program.

St. Louis Civic Improvement League

St. Louis's Civic Improvement League formally organized in 1901 out of a combination of existing clubs, including the white women's Wednesday Club.[34] The organization "did not try to usurp any of the functions of city officials," instead keeping to the sidelines and providing sanitary services the city did not. The two-thousand-member organization set up "breathing spots" on vacant lots, set aside for mothers and children to rest and take shade.[35] The league exerted some political pressure by printing a pamphlet, "Keep Our City Clean," which included text from garbage ordinances and encouraged citizens to complain of poor pickup to the garbage inspector.[36] The Sanitary Committee of the league, headed by a woman, suggested that women replace the police as inspectors.[37]

After the downfall of Ed Butler, the Civic Improvement League was ready to make a move on garbage collection. The league researched best practices of other cities, and in 1905, it published a forty-five-page report recommending garbage collection and disposal practices for St. Louis.[38] It claimed reduction to be the best form of disposal; it advocated collection by the city rather than contracting out; and it advised that household garbage items be separated. The league was initially pleased by signs of success. The city entered into a contract with a new reduction company, and the municipal assembly considered a bill regarding city collection. By 1909, however, the reduction plant proved incapable of handling the volume of incoming garbage, and the city collection bill died in committee.[39]

During the Butler era, the city's health commissioner reported that few citizens complied with instructions about where to place garbage cans. Garbage trucks could not pass through many alleys, but residents continued to place their garbage there. The garbage would not get picked up, and the contents of receptacles was inevitably scattered. In likely racially coded language, the administration denounced the "untidy and shiftless citizen" who treated alleys so poorly.[40] Civic-minded members of the league, by contrast, focused on model citizens. The residents actually chose to place their garbage receptacles in the alleys, they explained, as it kept them out of sight. When one woman tried to put her barrel in the alley but collectors would not pick it up, she was so dismayed at having to leave the "unsightly" barrel at the curb that she devised an enclosed bin to sit at the edge of her property (figure 5.6).[41]

The Civic Improvement League had its own ideas about sanitary practices, and it turned its attention to educating the citizenry.[42] This civic work of educating citizens operated within the Jim Crow policies that kept St. Louis segregated. In 1916 a popular referendum proposed an ordinance segregating the city through zoning. The NAACP took the ordinance to the Supreme Court, where it was declared unconstitutional.[43] Housing segregation made race and class disparities glaringly evident in the condition of homes and yards.

The Civic Improvement League encouraged clean yards, hosting a contest for most improved backyard.[44] The league organized to assist in enforcing sanitary law. The Ladies Sanitary Committee worked with a doctor who was appointed to investigate tenement houses. The committee hoped to educate residents on how important it was to use the correct garbage cans, and to show them the unsanitary effects of

Garbage bin erected by Mrs. Ella Hilton at the side of her house on Bell avenue, at Grand avenue.

FIGURE 5.6. Innovations in garbage can containment
Source: *St. Louis Republic*, October 28, 1900, 6, Missouri Historical Society, St. Louis.

improper household disposal.[45] They invited a speaker to discuss the sanitary challenges of urban life.[46] By 1910 the Civic League, as it was now called, made additional inroads with the City Council's Sanitary Committee, providing research for bills about tenement privy vaults and basements. Dr. L. E. Lehmberg of the committee was sure that these measures "would force compliance."[47] Despite these efforts, that same year the mayor took city officials on a tour of Chesley Island, the city's waste dump, where they found garbage piled two feet high and overrun with hogs.[48]

In 1911 the Civic League instituted a complaint bureau.[49] The Ladies Sanitary Committee sent a delegate to Chicago to learn the methods of an inspector there and urged St. Louis to hire one. The city did not, so the league's Sanitary Committee continued to focus on certain wards in sanitary need and appointed its own inspectors to visit wards and tenement houses while maintaining its relations with public officials such as the health commissioner and captain of the police department.[50] Women's organizations found that they could do their own outreach to low-income neighborhoods. They could teach sanitation.[51] This task made the most of underutilized middle-class women who could leverage their knowledge in aid of the working class.

Pittsburgh's Civic Club of Allegheny County

In Pittsburgh, women's groups were excluded from politics, largely because power resided in the Magee-Flinn machine, which was more concerned with protecting its control over the city rather than cleaning it. The Women's Health Protective Association (WHPA) of Pittsburgh was formed in 1889, with garbage as one of the "three universal and inescapable nuisances" that constituted its mission.[52] The WHPA took credit for the passage of the city's 1895 garbage ordinance,[53] and yet the organization had little to do with collection.

In 1895, faced with the dominance of the Magee-Flinn machine, the WHPA merged with the Twentieth Century Club to form the Civic Club of Allegheny County (CCAC). Women club members joined with men, including Allegheny's mayor William Kennedy as well as clergy, professors, and other progressive professionals. They organized against "prevailing municipal abuses" and for reform. At their first meeting, Kate Everest, the head of the Kingsley Settlement House, drew attention to the disparate effects of city policy on the poor. Others spoke up about sanitation. Women civic reformers had marshaled the resources of some of the civic and business leaders of Pittsburgh and Allegheny.[54] The mission of the CCAC was to promote "a higher public spirit, and a better social order" by holding public meetings, distributing printed information, and suggesting uplifting civic-oriented enterprises.[55] The club was positioned to challenge the Magee-Flinn regime, but the CCAC had little involvement with the administration.

Recourse to the judicial branch was no more fruitful. In May 1896, fifty-two residents of the neighborhood on the bluff above the reduction plant filed suit against Pittsburgh's trash disposal contractor, the American Reduction Company, claiming that the odor from the plant constituted a public nuisance. In response, the company altered some of its practices, making sure that the digesters were tightly sealed and that the pipe dispensing waste into the river was secured. The plant stopped producing fertilizer, instead shipping the raw material to a facility in the countryside. Residents continued to complain. This time the odor was found to be coming from the garbage as it sat in a pit awaiting processing. The Supreme Court of Pennsylvania determined that the residents had to put up with the inconvenience. While recognizing that the reduction plant presented a nuisance, the court determined that such nuisance was unavoidable. The bluff was once "probably the most delightful place about the city for residences," but the city had since grown and expanded, and "the onward march of the business of the city cannot be arrested because it may be an annoyance to some of those living on the bluff."[56]

There was no stopping Pittsburgh's machine, but in the coming years this coalition of reformers actively opposed it, achieving whatever results it could. Unable to get involved in the machine's garbage collection program, the CCAC turned its attention to a campaign to provide garbage cans in public places. The club identified the old bottles, wastepaper, scrap metal, and other refuse that was not included in household garbage collection and considered running a scrap resale operation.[57] The CCAC proved keen on scooping up garbage-related work that the city overlooked, carving out a place for itself.

By the early twentieth century, the CCAC was promoting the collection of garbage by the city in place of contracting out.[58] In the meantime, the club set itself the task of watchdog, recording the condition of yards and streets and sites of spitting and other such nuisances. The sanitary department of the CCAC documented, by date and place, incidents of unsanitary behaviors by city residents. Complaints included "expectorating in the openings between windows and backs of seats in closed street cars," backyards in "filthy condition," filth or overflowing garbage cans in playgrounds, open gullies exposing gas pipes, a grocer throwing fish brine and garbage into the sewer, and a manure box unemptied and overflowing.[59] Identifying garbage not picked up would come to naught because of the CCAC's difficult relationship with the political machine, but it had success with other complaints. After ten years "filled with alternate hopes and discouragements," the organization saw the passage of an anti-spitting ordinance and noted that the "little blue and white signs in the streets are monuments to the persistence of the Civic Club."[60]

White women's clubs and Black women's clubs did not work together in Pittsburgh. When a special train car was hired to take various club women to Lancaster for the State Federation of Pennsylvania Women meeting, the list of clubs was varied, but it did not include any Black women's organizations.[61] When asked if Black women's clubs would join white women's clubs at the state federation, Anna S. Posey, one of the first Black teachers at a white school in Ohio, countered, "Why should they? What would they gain?" Black women's clubs had a clear sense of their mission and boundaries. Posey referenced the mission of Black women's clubs, "They are climbing in a literary sense," invoking the motto of the National Association of Colored Women (NACW), "Lifting others as they climb."[62]

At home, the Black women's clubs of Pittsburgh operated—at least publicly—as literary societies. Musical, literary, art, and historical societies all provided a space for Black women to participate, where membership was distinct from that of white clubs.[63] Black civic associations gathering as literary societies dated back to the antebellum era. They were spaces that allowed for the "spreading of useful knowledge" by invited speakers and "free and full discussions."[64] Lecturers invited to Pittsburgh clubs over the years included W. E. B. Du Bois, Mary McLeod Bethune, and Langston Hughes.[65] Pittsburgh's So-Re-Lit (Social, Religious, and Literary) Club held fundraising dinners and garden parties for the Shadyside Hospital and "philanthropic homes." That work was summarized simply as philanthropic in the Works Progress Administration's *History of the Negro in Pittsburgh*. Black club women turned their efforts toward hospitals, especially in towns where Black patients were refused care. So too, those "philanthropic homes" may have been homes for the aged or for unwed mothers or for working women.[66] These efforts focused on community development and services that were not provided by local government.[67] Such work, then, was political in that it filled in the gaps for underserved Black residents, though separate from the more clearly political activity of the National Association for the Advancement of Colored People (NAACP) and other organizations.[68] One such organization, the Frances Harper League, passed a resolution urging local stores to hire Black men and women to be shopkeepers, adding that the customers at their counters were both Black and white.[69] When these clubs met together at the State Federation of Negro Women Clubs in 1911, the state organization denounced a lynching that had occurred in Pennsylvania, making the point that such lawlessness as had been happening in the South was happening as well in their own north-ern state.[70] In 1912, association members denounced lynching and Jim Crow in the same speeches in which they promoted improvement in the home.[71] Members' interest in sanitary matters was acute and immanent. Given the identification of high Black mortality rates in board of health reports, the association resolved that "the Negro is not subject to tuberculosis any more than any other race but that it is the unsanitary conditions under which thousands of them are forced to live" that was the cause.[72]

Pittsburgh's industries needed labor, bringing in immigrants and workers who migrated from the South. Not much new housing was being built, so new workers crowded into existing accommodations. In 1914 Helen Tucker had already noted the poor condition of hous-ing for Pittsburgh's Black residents, who crammed into alleys in the Hill District or in substandard housing near the mills. She observed that garbage was picked up only once every other week. Those who could afford to left for other neighborhoods.[73] Although migration increased—dramatically—in 1916, no new houses were added. That meant that new arrivals were packed into attics and cellars, store-rooms, basements, and sheds. One observer noted, "The conditions in rooming houses often beggar description." This overcrowding meant poor sanitation and illness. A cooperative effort between the bureau of sanitation, physicians, and Black organizations carried out an educa-tional campaign to address issues ranging from ventilation and care of infants to the dangers of patent medicine and carelessness in dress.[74] Addressing the living conditions in crowded spaces was one of the resolutions passed in 1896 by the NACW. With regard to families liv-ing in one room, the NACW urged that mothers' meetings be used to teach mothers "the necessity of pure homes, and lives, and privacy in home apartments."[75]

A Russell Sage Foundation-sponsored report, *The Pittsburgh Survey*, identified leading civic organizations including Kingsley House and the Civic Club of Allegheny County.[76] Kingsley House was a settle-ment house devoted to investigating social problems and meeting the needs of the poor in Pittsburgh's neediest neighborhoods, namely, the Twelfth Ward and the Hill District. The *Survey* had headquartered itself there while its research was being conducted.[77] Kingsley House workers documented the conditions in these neighborhoods, replete with over-flowing garbage cans (figure 5.7) and carts (figure 5.8), with poignant indications of children living in the midst of squalor.

FIGURE 5.7. The Kingsley Association documented living conditions in Pittsburgh
Source: Kingsley Association Records, 1894–1980, AIS, 1970.05, Archives Service Center, University of Pittsburgh.

FIGURE 5.8. The Kingsley Association documented uncovered garbage carts
Source: Kingsley Association Records, 1894–1980, AIS, 1970.05, Archives Service Center, University of Pittsburgh.

Model Housewives

Pointing to neighborhoods in need could direct services to them, or it could distinguish residents by their class, race, and citizenship status. When city departments needed to account for their lax performance in collecting garbage, city officials would routinely report that the department was doing fine work, and garbage collection would be a success but for noncompliance, particularly in those neighborhoods.

Women's Civic Organizations in Pittsburgh

Pittsburgh's South Side and other neighborhoods continued to suffer from poverty and inadequate living conditions. Residents complained that their trash pickup was irregular and that collectors would toss the trash in "the most convenient spots." The collectors claimed that they visited daily, but since garbage was not left in the proper receptacles, it was difficult for them to collect.[78] Despite receiving instructions, householders failed to follow them.[79] Living and sanitary conditions had been a concern for decades, but it was not dealt with through reform. By the 1910s, the poor neighborhoods of Pittsburgh had become an object of reformers' studies and complaints. When Pittsburgh city officials did not succeed in collecting the garbage in 1916, they deflected attention onto the housing available to poor people, calling it unfit for human habitation.[80]

While Black women's clubs, keeping a low profile, were attentive to the conditions of the working class, white women's clubs cast residents in constructed identities and exposed their private conditions as public problems. In publicizing substandard living conditions, middle-class white women likewise constructed a public identity for themselves. The dynamics are illustrated in an anecdote in the 1908 *Pittsburgh Gazette Times*. A garbage collector emptied only one of three cans left outside a home. The house's resident was watching through the window and went after him immediately. When pressed, he said he would empty the other two next time. The woman countered that the cans had been there for weeks, and she threatened to fetch a policeman. The collector figured she was just bluffing, but she went inside, put on her bonnet, and marched down the street to find a police officer. When the collector realized she might actually do it, he went back and emptied the rest of her pails.[81]

This anecdote relies on a number of racial tropes. The collector is described as a "husky colored man," and in the sketch accompanying the article, he looms over the diminutive but determined woman. He speaks in dialect, while the white woman issues a sharp institutional warning that she will use the policeman as a witness in appealing to the authorities. If the story did not get the point across, the illustration summed up the power of the small white woman against a larger Black man. The white woman, shut out of formal politics, could nevertheless enlist the threat of law enforcement to wield power in this sphere, or she could have a "change of heart" if the garbage collector complied with her demands. Contrast this to Black club women who resolved to tend to the home, which included addressing sanitary conditions and preventing the home from becoming a conduit for the police. The home was not an equivalent space for white and Black women. As Rosetta Douglass-Sprague, the daughter of Frederick Douglass, put it, speaking on the first day of the first meeting of the NACW, "We want homes in which purity can be taught, not hovels that are police-court feeders."[82]

Although both white women's clubs and Black women's clubs in Pittsburgh were pushed out of formal governing, each of these groups displayed political consciousness. White women's clubs had more access to the opposition party and to civic leaders. They were able to play a part in coalitions and to carve out a space as collectors of complaints—whether about spitting on streetcars or the performance of garbage collectors—giving themselves the job of surveilling social mores related to sanitation. Black women's clubs promoted sanitation through fundraisers and under the guise of being literary societies, while white women's organizations developed their power in the civic sphere to surveil and render judgment—on residents, garbage collectors, and municipal government.

Louisville Civic Association

Unlike in Charleston, St. Louis, and Pittsburgh, Louisville's women's organizations enjoyed access to city offices that allowed them to disseminate information and network with public officials. Louisville had public trash collection, housed in the department of public works.[83] City officials were in touch with leading sanitarians of the day to consider innovative options, although they did not follow up on them. The mayor and public works officials met with the sanitarian Rudolph Hering in 1909 to discuss advances in garbage incinerators.[84] But in 1917 he and Samuel Greeley reported that the city continued to use the less advanced method of a dump, although it was well kept.[85]

As in the other cities, women's civic organizations in Louisville made inroads in supplementing the city's household garbage collection program. During Louisville's two-day "Clean-Up Days" in 1912, the city devoted twelve wagons for the street department to haul away 2,324 loads of dirt and trash. The event was pronounced a success.[86] Louisville officials allowed women's clubs to take an increased role in garbage collection through public programs using women's clubs as an extension of municipal authority. When complaints were raised about garbage collectors missing their rounds, or uncovered wagons, or conditions at the dump, the Women's Civic Association looked into them. With their own dedicated investigating committee, members followed garbage wagons in cars, observed collection and disposal practices, and then called on the chairman of the department of public works to present their views on needed reforms.[87] A member reported that the association had found not only poor public practices but also ignorance and carelessness about the garbage issue among the population.[88]

The association had access to various networks, working with other women's clubs and sanitarians, and to public officials in their city. In 1915 the Louisville Women's Civic Association invited Mary McDowell, renowned for her work in improving Chicago's garbage collection, to speak at a mass meeting. The audience included members of the board of public works, the city health officer, and other city officials.[89] In 1917 James Caldwell, chairman of the board of public works, delivered an address to two hundred women at their club meeting. Samuel Greeley, consulting sanitary engineer in Chicago, talked to the group as well.[90] The civic association exhibited photos exposing substandard trash conditions. It printed a pamphlet, "Facts on Louisville's Garbage Problem," and distributed four thousand copies. To further reach women householders, the association printed up notices headed "Housekeepers, Attention" and distributed them through schoolchildren. In this way they were able to inform fifty thousand households to keep their garbage cans covered.[91]

The organization even produced a movie, *The Invisible Peril*, dramatizing the dangers of unsanitary practices of garbage collection and disposal. Funded with support from the Board of Trade, the Rotary Club, the Men's Federation, and the Jefferson County Medical Association, the movie followed the trail of a discarded hat as it traveled from open can to open wagon to an open dump. The "dump picker" who finds the hat ends up in the sickroom. The movie reached an audience of over twenty thousand people—both government officials and the general public.[92]

Louisville's population reached 204,731 by 1900. The sudden increase in population in the late nineteenth century was met with racially segregated neighborhoods.[93] Although the city did not formally disenfranchise the Black population, political machines relied on police intimidation of Black voters.[94] The segregation that had developed informally began to be formalized, with a proposed ordinance prohibiting Black purchasers from buying houses on blocks where white residents lived, and vice versa. The NAACP challenged the ordinance, and it was declared unconstitutional by the US Supreme Court in 1917, citing the freedom to both use and acquire property under the Due Process Clause of the Fourteenth Amendment and the civil rights intended to be protected by that amendment.[95]

In response to declining industrial development in the first decades of the twentieth century, the Board of Trade launched the Million Dollar Factory Fund to foster creation of small businesses in 1916.[96] In the context of this plan to revive the local industrial economy, an ordinance in 1917 required that every householder, restaurant operator, or hotel-keeper, and all keepers of businesses, use watertight, fly-proof receptacles for garbage and place ashes in a separate container. The city, in turn, was required to adopt the use of watertight wagons and dispose of garbage in such a way that it would not endanger public health.[97] When the Board of Public Works instituted a scientific garbage collection program in one section of the city with federal funds, the Louisville Women's Civic Association partnered with the Board of Public Works and the US Health Service to launch a program to separate wet from dry garbage and feed the wet garbage to hogs.[98] Major L. D. Fricks of the US Public Health Service directed remarks to housekeepers, chiding them for failing to comply with the separation program.[99] His program was supplemented by a "Kitchen Card" laying out the rules for the housekeeper. Garbage collection service was extended to the Point District, which was "absolutely without a garbage disposal system," according to a survey conducted by the Women's City Club.[100]

White women's civic organizations were able to offer their resources in this reconfiguring of Louisville's racial and economic conditions. They did so by showing that they could mediate between experts and common people, networking with experts and then educating the masses. Their middle-class status served them well in being seen as models of correct behavior.

In the 1930s and 1940s Louisville invested in technological improvements. The city bought a new sand spreader, which would make streets less slick, and replaced "old time brooms with a mechanized 'Panzer' for a 'blitzkrieg'" in its street cleaning equipment, but it was losing the war on filth.[101] Louisville ran up against the old problems of residents improperly sorting and putting out their garbage for pickup. In 1941 the city's finance director sought to induce citizens to buy proper garbage cans with covers that would stay on tight. The program fell through, as not enough cans were available.[102] Rather than turn to the voluntarism of civic associations, this time the city added women to the payroll. Three women were hired to work under the superintendent of sanitation to reach out to housewives and instruct them in the proper use of garbage cans. Residents needed to be told how to separate garbage and wrap combustible material. The sanitation visitors were sympathetic, acknowledging that women were busy. Rather than talk down to them, they tried to reach out to them and let them know that uncovered, unsorted garbage attracted flies, rats (another public problem the city was trying to address systematically), and bugs, all dangers to children.[103]

Birmingham "Overseers"

Birmingham, too, invited women's clubs to participate in addressing sanitary problem. The "Pittsburgh of the South" was a relatively wealthy planned city, incorporated in 1871 as a site of manufacturing and mining. Technically, Birmingham contracted out its trash collection and disposal efforts.[104] As early as 1890, city boosters touted the twenty miles of Waring system sewerage and street cleaning and garbage collection practices.[105] But by 1910, both its garbage collection and sewerage were inadequate at best, and dangerous at worst: 10 percent of the city used outdoor privies, and death rates from typhoid and tuberculosis were high. The collection wagons deposited their filth in open lots or in dumps, where sometimes it was well covered by ashes but other times left open to be picked over by cows and pigs.[106]

From the outset, the city relied on Black labor while segregating Black and white residents in different, unequally serviced neighborhoods. Areas with predominantly Black residents lacked indoor plumbing and running water.[107] Such conditions persisted until the mid-twentieth century. At the start of the New Deal, 30 percent of Black homes in Birmingham lacked sewer connections; garbage was not collected; and streets were unlikely to be paved in Black neighborhoods.[108] As late as the 1960s, Black communities still sought to secure basic municipal services, including sewers, paved streets, and emergency services.[109] When, in 1916, the *Birmingham News* called upon Black civic leaders to teach residents about proper garbage disposal, the NAACP magazine *The Crisis* wondered, "What in the name of common sense can poor, disenfranchised laborers do in the matter of sewage and garbage disposal in a city like Birmingham?"[110] The answer, *The Crisis* responded, was to leave. Workers were migrating to better conditions in northern cities and leaving behind deplorable neighborhood conditions.

Black civic organizations drawn from the middle-class community responded to the disparities. For example, the Climbers, a club for women teachers, raised funds for a "Home for The Aged and Destitute Negroes," an anti-tuberculosis club, and a "social uplift" campaign.[111] It was the white women's clubs, though, that were invited to help out officially. The mayor gave white women's clubs, organized by city block, supervision over street cleaning and the disposal of waste and garbage. As Rheta Childe Dorr remarked in her survey of women's clubs, "They really act as overseers, and can remove lazy and incompetent employees."[112]

Though the women's clubs in Birmingham resembled those in Louisville, their role in administrative oversight was quite different. Women were more like official supervisors, while the men over whom these women's block clubs served as "overseers" were very likely Black, and possibly convict labor. Birmingham (like New Orleans and Charleston) maintained a convict labor force for decades. Vigorously enforcing vagrancy laws allowed police to discipline the Black population while providing a steady stream of labor for the mines that served as the basis for Birmingham's economy.[113] Convicts were put to work on city projects too. The street department used a convict force.[114] In 1915, when the city drastically cut back on municipal services from police and fire to schools and libraries, of the one hundred men employed to collect garbage, forty-two were convicts.[115]

While white supremacy was maintained by protecting white women from Black men, they met routinely, and mundanely, when Black men came to white women's homes to pick up garbage.[116] In those situations, white women could wield their privilege, either as the model consumer, there to be served, or as "overseers," subverting the cultural identity of the vulnerable woman in need of protection and asserting her race and class privilege. Just as municipal housekeeping extended women's sphere from the home to the streets, white women's privilege could be extended into the public sphere as well.

There's a Right Way

The housewife's compliance and model behavior are epitomized in the *Louisville Courier-Journal*'s September 28, 1941, article "There's a Right Way to Do a Thing . . ." In the accompanying photo, a white woman stands in her driveway, watching two Black collectors picking up her garbage (figure 5.9).

Mrs. C.O. Ewing, according to the caption, "has a sanitary container, has wrapped her combustible waste in paper and separated it from non-combustible material—all required by city regulations." Furthermore, she puts her combustible material out one day and the noncombustibles the next. There is no need to be confused: "The garbage man can tell you which one."[117] Mrs. Ewing distinguished herself from housewives who openly "flouted" the rule—those housewives who complained that

FIGURE 5.9. Assuming the role of model citizen
Source: "There's a Right Way to Do a Thing . . . ," *Louisville Courier-Journal*, September 28, 1941, 69, Louisville Free Public Library.

their garbage cans were frequently stolen, that they were too busy to follow the rules, that they had always taken care of their trash in their (substandard) way and were not amenable to change.[118] Mrs. Ewing took advantage of her class and race to position herself as both moral authority and capable administrator.

Louisville's model housewife was on good terms with her garbage collector. The explicit supervision of Black garbage collectors, whether informal or formal, by white women in Louisville and Birmingham showcases the racial hierarchy on which white women stood when being cordial with and paternalistic toward the collector who visited her driveway. The relation between housewife and garbage collector indicates that there was another habit that the state needed to cultivate. By granting white women the privileges of their racial status, the state enabled them to assert their dominance over Black collectors approaching the white woman close to her home.

Gender Hierarchy as Resource

The garbage can problem was a problem of implementation, but the actors who would do the implementing were private actors, starting from the intimate physical space of their own homes. The garbage can problem's solution exceeded the standard public-private relationship. It revealed just how much power the state needed citizens to exercise in their own homes. The state wished to elicit behavior. To make garbage collection work, week after week, the state had to go even further and inculcate habits in residents so as to internalize appropriate behavior in them. The state did not need to draw on racial and sexual identity, but it would help if residents were to pride themselves on being the kind of people who disposed of garbage appropriately. Proper garbage can practices could become part of one's identity, and appeals to racial and gender status were means for shaping it. Women's organizations were an available resource for the molding of identity. Women were able to craft their own identity as political actors, and in turn they were able to raise their own status as ideal civic actors with authority to model behavior for the poor and persons of color.

Cities' experience with the garbage can problem indicates why a city might construct racial identities to further government aims. The garbage can problem required that public authority intrude into private space. Not only that, but to ensure ongoing compliance, citizens had to change their habits, which required an exercise of state power. Done despotically, the process could be heavy-handed, and even lead to the end of an administration (as in New Orleans). Infrastructural power instead invited citizens to shape themselves into the sort of person who complies with proper garbage collection. Structural problems became matters of individual merit or failing. Shaping the right sorts of people requires engagement with identity. There are numerous options in choosing identity, but race, class, and gender remain prominent, and they are convenient for the state to use, as they operate in the social realm for public purposes.

The cleavages in status continue to be readily visible as communities clean up their garbage and rely on available resources to educate and model appropriate behavior. This suggests that status is not a historical vestige to be overcome by political development. Rather, it is a tool enlisted by the state to carry out its work, and it is incorporated into public programs.

Cities across the country faced a similar garbage problem at the same point in time. Although their specific solutions differed, all seven we examined drew on infrastructural power to inculcate the proper sanitary habits. On the one hand, this form of power created good citizens, model citizens who put their rubbish out at the right time and in the right place for collection. On the other hand, in contrast to model citizens were the lazy, dirty people who were cast as lacking basic sanitary skills and understanding. These people were held responsible for the problem of noncompliance with city ordinances (generally) as well as for their own poor health and high mortality (specifically). Municipal trash programs, which drew on infrastructural power, created policies that allowed some residents to take credit for their moral deservingness at the same time they blamed failures on others. The state employed race and gender in early sanitation policies, institutionalizing credit and blame into state building.

Women had been pushed out of municipal garbage collection when corrupt municipal regimes secured control over the operation and benefits of garbage collection. Women were available as a resource, however, when those city governments ran into problems with householder compliance. They could simply have distinguished themselves as model housewives, buying the latest garbage can and resting on social status to elevate themselves. But the solution to the garbage problem was not relegated to the private sphere. Cities fostered relations with civic groups and enlisted women in governing. Birmingham and Louisville provided women's civic organizations with the most access, highlighting the dynamics of power between white women and Black city workers. When we apply those dynamics to the other cities, we can see those relationships in the private sphere, where a white housewife could wield her privilege over the Black workers who entered her space. White women were able to muster a proprietary concern for community sanitation and position themselves as experts and model citizens over others.

That reliance on gender drew on and constructed stratifications of citizenship based on gender, race, and class. Gender hierarchy was a resource that a government could tap into to get its work done, and that reliance, in turn, exacerbated status categories. In relying on status, the government altered it and incorporated those new status categories into modernized political development. This incorporation of status is particularly evident in the women's groups that were hired, or that acted, as "overseers." But women did not need to be formally employed in the process to engage in it. A white housewife lecturing Black men in her driveway, instructing them on the "right way to do a thing," showed that women did not need formal employment to play a role that allowed them to place themselves above others. Even in Pittsburgh, where women's groups lacked such access, the white woman threatening to call the police on the garbage collector illustrates the power that middle-class women could muster because of their claim to know more than others and to instruct them.

CHAPTER 6

Getting and Keeping Garbage Collection
Municipal Reliance on Racial Hierarchy

Municipal governments built capacity for trash collection and disposal with boards of health, street departments, boards of public works, contractors, horses, carts, and crews of drivers and collectors. City officials fostered public-private relations to induce residents' compliance. Still, into the early 1900s, municipal garbage collection continued to be underdelivered or spotty. When that became apparent, public officials would cast about for someone to blame and usually landed upon residents, painted as uneducated, or garbage collectors, cast as lazy. Quite often, this blaming was inflected with race or class.

Certainly race enters the story of garbage collection in terms of both who collected trash (immigrants and people of color) and whose neighborhoods were poorly served (the same). Yet race plays another important role. Racial and class hierarchies were resources that municipal governments could rely on to get and keep garbage collection by constructing stories about the shortcomings of Others. By invoking racial hierarchy, local governments incorporated race and class disparities into the political process and rendered them tools of state building.[1] Modernization of institutions does not necessarily bring with it modernization of ideas, if those ideas of inequality can help a regime politically as it musters resources. Across the three cities we examine in this chapter—Pittsburgh, Charleston, and San Francisco—racial hierarchies were used in a bid to enhance power, employed both when one interest tried to secure garbage collection from another and in deflecting criticism once a collection program was up and running.

Racial Disparity in Garbage Collection

Racial disparities pervade the history of garbage collection in America, reflecting larger patterns of systemic inequality. Neighborhoods were served or underserved according to class and racial lines, a situation publicized by activists in the early twentieth century. The Neighborhood Union of Atlanta, started by local women, drew attention to unsanitary conditions in the city's Black neighborhoods, pointing to disparate garbage collection as well as the burning of garbage in the vicinity.[2] Garbage collectors were often Black or immigrant men and were frequently ignored until they went on strike, bringing attention to the essential services they provided.[3] The Reverend Dr. Martin Luther King Jr. was assassinated in Memphis, where he was visiting that city's 1,300 striking garbage workers. They were contesting the practices of white department of public works officials toward predominantly Black sanitation workers.[4] Racial inequality was present at the outset and accompanied the development of garbage collection. Yet race itself is barely evident in official city reports about garbage collection, emerging only at particular moments. When regimes deflected criticism, racialized groups were distinguished in a practice of scapegoating.

We find the occasional appearance of race in official public records, in which race is a latent but potent resource of governance. At the outset of municipal garbage collection, Pittsburgh was a predominantly white city but with increasing numbers of European immigrants and Black residents. Charleston's municipal collection operated in a city in which the Black community constituted 56 percent of the total population.[5] After Reconstruction and into the Jim Crow era, the city relied on some of its antebellum tools of racialized social ordering and white supremacy to secure white dominance. San Francisco was a predominantly white city that was classified as having no municipal collection because it relied on private scavengers. Still, its patterns of invoking race match those of the other cities. References to race could be absent for a period of time, only to be called upon when a regime or an agency or organization needed to maneuver through a political challenge. Regimes that integrated corruption and had a hold on power largely avoided racial references. It was up-and-comers, or sitting organizations that were challenged, that called up racial differences and connotations. Relying on race indicates vulnerability in the regime, which invited racial hierarchy into institutional development.

This maneuver was not new, as it was a significant tool in claiming ownership of garbage collection in the first place. Prior to citywide collection, there had always been some entrepreneurial individuals who would cart away garbage. When local officials wanted to take over garbage collection, they would invoke familiar racial tropes to wrest control from the scavengers. Before cities had formal collection programs, these entrepreneurs often operated by approaching individual households and businesses and offering to haul away trash. Although the arrangement was informal, scavengers were a mainstay in municipalities and were even relied upon by city agencies to deal with cleanup projects. Lacking the capacity to collect garbage, local officials decided that scavenging was the best they could do. When regimes found they could benefit from a citywide program, they had to get rid of the scavengers.

If a city needed to wrest control from scavengers, then racial hierarchy was a resource for casting aspersions on the unregulated and presumably unsanitary habits of these individuals. Pittsburgh officials were ready to take this course of action, while Charleston officials kept relatively mum. This is where the earlier integration of Charleston's corrupt regime with city government is evident. Charleston's regime had less need to cast aspersions, as it already had close control over the use of scavengers through relations with slaveholders and, later, racialized social ordering. Pittsburgh's sanitarians, by contrast, had already moved toward modern collection practices prior to the rise of the Magee-Flinn machine. Lacking political clout, they resorted to racist and classist insinuations.

The surprise is San Francisco, whose entrepreneurial scavengers appeared by the 1870s and never lost control of collection to the city. The city periodically challenged the scavengers, who held on to their dominance through various tools of collective action and resistance. One tool was their construction of the specter of a racialized Other in their successful grip on garbage collection.

Wresting Collection from Scavengers

When disposal of garbage was still a household responsibility, residents generally took care of their own waste, however they could. They might throw ashes in their backyard. They might feed their kitchen garbage to chickens or pigs.[6] Prior to municipal garbage collection programs, the task was accomplished by "a collection of trades, each catering to a different part of the waste stream."[7] Some cities had herds of swine roaming the streets. Scavengers took trash off the hands of householders, and they eked out a living from their entrepreneurial efforts. A scavenger would make an arrangement with a household or business to haul away their garbage. When cities were small, scavenging was manageable; the scavenger could take the garbage to nearby farms for feed. Or he might just dump it in an empty lot or in the river. Under this system, across the country there was little incentive for the householder to pay for clean or efficient disposal, and little reason for the scavenger either. Nor was there much motivation for householders to comply with expectations of keeping their homes or curbs clean. They did as much or as little as they pleased.[8] While not ideal, it was an arrangement that could be tolerated when cities were smaller and housing was less concentrated.

Seizing Control in Pittsburgh

A city might catch infractions of sanitary ordinances through nuisance laws. In 1807 Pittsburgh fined people who disposed of animal carcasses, garbage, noxious liquids, or other offensive matter on any street, square, lane, or alley, or along the rivers, or within an enclosure.[9] In 1848 the city targeted anyone who threw garbage or any sort of filth into the canal with a fine of $20 and made them responsible for removing it.[10]

Nuisance laws, however, could not actually keep cities clean. Health concerns raised the perceived need for more governmental intervention to ensure better garbage removal. A cholera outbreak in 1832 was recognized as a public problem, and the city devoted resources to street cleaning and garbage removal.[11] Pittsburgh's establishment of a board of health in 1851 gave it authority "to have all objects which may have a tendency to endanger the health of the citizens removed or corrected as [it] shall deem necessary for the health of the citizens." It had the authority to declare nuisances, for which violators would be fined.[12] The board members also had the authority to "to employ, from time to time, as many scavengers as they may deem necessary, upon such terms and with such appliances and conveyances as they may deem expedient, and to make, from time to time, such rules and regulations for the conduct of such scavengers, as they may deem necessary."[13] At such times the city might enlist scavengers to help, but scavengers did not provide the capacity for extensive collection. They instead concentrated their efforts in commercial districts and, according to Lawrence Larsen, the profession "did not draw the highest type of individual."[14] In the late decades of the nineteenth century, scavengers were an expected resource in garbage collection. The board of health could give notice to hotels, taverns, eating houses, and private houses that the scavenger would come and that the owners should have their garbage ready. For their part, the scavengers were required to keep their wagons cleaned, disinfected, loaded properly, and covered so as not to spew garbage through the streets.[15] Scavengers provided capacity to a board of health that had no mechanisms of its own for picking up the garbage.

In 1875 the board asked for authority to remove street dirt and clean the sewers, citing the city's high infant mortality rate of 53 percent for children under the age of five.[16] Pittsburgh, though, had a high tolerance for dirt. After all, in an industrial city, smoke in the air was an indicator of prosperity as much as it was a nuisance.[17] The board of health was granted the authority it requested, but it was given little in the way of capacity to clean or enforcement powers to punish violations. This changed in 1880, when typhoid fever struck, surpassing the wrath of yellow fever in Memphis. The northern city of Pittsburgh was susceptible to the epidemic diseases that so regularly occurred in warmer southern climates. Pittsburgh's low-lying South Side neighborhood was poorly drained, and it was hit particularly hard. The board of health reported nine new cases a day, laying the blame on "the constantly arising effluvia from badly constructed and reeking sewers."[18] Finding that neighborhoods with high rates of typhoid fever also suffered from high rates of diphtheria, the board recommended that the sewers, streets, and alleys be cleaned. It licensed a boat to remove garbage from the city.[19] Although the boat owner was given free wharfage for five years, the fee he charged was "enormous."[20] Citing a conversation with a man who found a pile of trash sixteen feet deep dumped in a vacant lot,[21] the board drew connections between the system of garbage removal and street sweeping and the spread of infectious disease. The board asked for control of garbage removal.

The typhoid epidemic provided an entry point for a board of health that had been trying for years to secure capacity and enforcement powers. The board had been collecting mortality statistics and could identify the South Side neighborhood, which was already recognizable as working class, as a discernible site of disease. The board thus linked infrastructure to the epidemic and singled out a poor district as harboring conditions for the spread of infection. Targeting a poor neighborhood helped illuminate the need for public health measures, but it also vilified poor people as unhygienic and even a threat to the community. This tactic was again deployed when health experts tried to wrest garbage collection from the current, informal efforts of scavengers. Picking through garbage looking for items to sell to companies that would reuse them was a product of industrial development. With increasing amounts of waste lying about, and manufacturers willing to pay for reusable scraps of cloth or metal, trash was a commodity. Enterprising actors—often poor—would take advantage of the opportunity.[22]

As health experts sought systematic city collection, they turned their attention to the "ragpickers," hoping to replace them with a professional and disciplined force of collectors. In 1875 the board of health called a special meeting to take steps to "contribute to greater cleanliness and purity of the streets, avenues, alleys, sewers and byplaces in our city."[23] The meeting ended with a resolution to appoint a committee of five to prepare a report. Health experts were ready to plan for city sanitation in earnest.

It was at about this time that the scavengers came under public scrutiny. Scavenging was a dirty business, and scavengers were relied upon and looked down upon at the same time. Toleration of their activities could turn to derision when someone wanted to take over their job. And since they were likely to be immigrants or people of color, their status as well as their work rendered them easily subject to disdain.[24] According to an article in the *Pittsburgh Post* in 1880, "hundreds of men, women, and children—roam the streets calling out 'Rags!, Iron!' and pay a few cents for each pound collected and then will in turn try to resell them for profit." They lived in dark, dingy, dusty apartments, filled with heaps of smelly rags. After going out scavenging all morning, the reporter continued, "they then return with their load, and after dinner the 'sorting' is commenced. This is a most tedious job, and always lasts half a day." The reporter got the scoop on what things looked like in the home of one ragpicker, a Mr. Reymer:

> All the rags are emptied out of the bags upon the floor, and Mr. Reymer then seats himself amongst them. Each bit of rag, no matter how large or small, is picked, examined and thrown into its respective heap. There are three heaps, hard, soft and cotton rags. After all that was gathered in the morning are sorted, they are

bundled up and carted to the warehouses of paper firms. From there they go to the paper mills, and are finally returned to us in the shape of clean, white paper.... [The same process was] carried on by hundreds of men, boys and children in our city.[25]

Those who did the dirty work could be cast as a menace to public health. Health experts were asserting the need for the city to assume responsibility for sanitation in poor neighborhoods and to take over the work from scavengers. The board of health's report on the South Side urged that the city councilors pass an ordinance for "a systematic method providing for the thorough and regular removal of all ashes, garbage and offal under supervision of the Board of Health."[26] The board did not get control over effective garbage collection, but it was joined by new allies in the battle for public health. As the city's population increased and conditions of overcrowding and industrialization worsened, middle-class women took on public health as their cause. Through the efforts of the Women's Health Protective Association, women reformers submitted their own proposal for a garbage collection ordinance.[27]

Transforming Public Scavenging in Charleston

Whereas Pittsburgh's board of health relied on racial hierarchy to cast aspersions on private scavengers, Charleston directly hired scavengers. Racial hierarchy was involved here, too, because the first scavengers were enslaved persons, who were hired out to the city. Charleston continued to use scavengers in its later municipal collection. Charleston had one class of scavengers, the poor who picked through garbage for scrap, called "chiffoniers," like their Parisian counterparts (*chiffon* being the French word for "rag"). Chiffoniers were looked down upon, described explicitly as Black people "who ply their vocation from early morn to dewy eve. They are of both sexes, generally of an advanced age, and are to be distinguished by a chronic stoop." The *Charleston Daily News* referred to them as a "humble branch." Working without a horse or cart, a scavenger might be a Black woman who walked along with a canvas bag, scratching at dirt on the ground with a hoop of bent iron, hoping to find iron and rags to sell at the junkshops.[28] When they did use carts, scavengers could be noted for their "odious" contents—putrid meat, decaying fruit, greasy remnants—casting those odors about as the cart traversed the city streets.[29] After the Civil War, some sporadic public aspersions were cast at scavengers. A brief note in 1866, titled "Lazy Man's Load," pointed to the "carelessness" of scavengers in loading their carts and the "most offensive nuisance" that resulted from the contents spilling out on some of the city's finer streets. An ordinance for the better regulation of the street department followed in May 1868.[30] Yet people were still complaining in 1869. A letter writer to the *Charleston Daily News* revived concerns about the lack of sanitation of scavengers' carts, urging that the scavenger's collection of offal be curbed so as not to spread yellow fever. That letter writer suggested that the practice of using garbage as fill on lowlands should be stopped as well. Those solo scavengers could be replaced by city scavengers.[31] Accordingly, in 1871 an ordinance for keeping the streets and lots of the city clean required owners and occupants to sweep their premises, and the sweepings to be carted away by city scavengers.[32]

Soon enough, Charleston's sanitarians had found a foothold in government. In 1879 the Broad Street Ring's William Courtenay won the mayor's race, beginning an era of stability for the conservative Democrats. The board of health was reorganized, and in the 1880s health officials tried to claim its authority and do the job properly. Dr. Henry Horlbeck, who would serve as health officer for twenty years, was active in the American Public Health Association, attending yearly meetings and printing conference proceedings in his annual report to the mayor so as to promulgate news from sanitarians at the national level. He served as president of the association in the 1897 term. Charleston's health officials followed the progress of George Waring's sewerage program in Memphis and hoped for a comprehensive plan of their own.[33] In 1883 the health department spoke of inviting Waring to visit Charleston to recommend a plan for dealing with city sewage and waste.[34] Unlike in Pittsburgh, where sanitarians were pushed out of the garbage contracting process, Charleston's health officials maintained their hold on garbage collection. Responsibility for the city collection was shared between the street department—which provided the horses, carts, and drivers—and the health department, which provided a fleet of four sanitary inspectors who visited fifty premises a day, looking for violations of sanitary ordinances, including those dealing with garbage.

Maintaining Collection

Charleston had municipal collection much earlier than Pittsburgh, and its board of health was better positioned to regulate it. Apart from a few thinly veiled racial references during Reconstruction, Charleston's public records from that time ignored race. That changed as cities operated more modern garbage collection after the 1890s. As ruling regimes evaded responsibility for their own shortcomings, racial hierarchy became a resource to deflect blame in both Pittsburgh and Charleston. In San Francisco, immigrant scavengers themselves played on fear of the Other to keep garbage collection and disposal out of the city's hands and maintain their own services.

Deflecting Complaints in Pittsburgh

When the American Reduction Company received the garbage contract, it hired drivers and collectors who were predominantly—if not all—Black. The race of employees is not mentioned in the annual reports, but it was widely known.[35] Driver was one of the few occupations that Black men in Pittsburgh were able to claim and advance in. The *Pittsburgh Survey* reported that there were at least a dozen men who owned their own horses and wagons and took in hauling contracts. The largest outfit employed 135 men. Another had thirty employees for hauling while employing one hundred to two hundred men for paving. Smaller operators worked on their own and made a living. Half of the nine thousand teamsters in 1907–1908 were Black. In 1913 the International Brotherhood of Teamsters, Chauffeurs, Stablemen, and Stablemen's Helpers organized their own local. That same year Pittsburgh's garbage collectors went on strike, increasing their wages from $2 to $2.25 a day.[36] They went on strike again in 1917, this time in response to racial disparities. When Black workers at the Allegheny Garbage Company refused to sign a wage scale agreement, their places were filled by white workers. In solidarity, the three hundred Pittsburgh collectors, all of whom were Black, went on strike. The garbage was mounting, and as warm weather loomed, residents noticed.[37]

The collective action of garbage collectors was referred to with nods to their intemperance. What did Prohibition have to do with garbage collection? No longer able to slip into the saloon for "his little 'nip,'" the collector was allegedly less motivated to do his work, and the piles of trash were growing. "Booze and garbage go hand in hand, so it seems," charged the *Pittsburgh Daily Post.* The evidence of laggard service was provided by women, who were left out of any role in collection but kept a keen eye on it: "One woman said she hadn't seen a garbage collector since Thanksgiving. Another one said she hoped they'd clean up around Christmas."[38] While the race of collectors was hardly referenced at all in city reports, it came up whenever garbage collection became a problem, in this case because of workers going on strike. The response was often to draw attention to their race and imply drinking on the job.

Pittsburgh's collectors were required to wear a badge.[39] This was different from Charleston's disciplinary measures, but it still reflected residents' unease with their haulers. There were reports of random men posing as garbage collectors, approaching households for holiday tips.[40] The American Reduction Company did not offer stellar collection, but collectors themselves largely evaded being the target of complaints, except when the stakes were raised. When a family in a middle-class neighborhood was devastated by scarlet fever, eyes turned toward the two garbage collectors, who regularly left their loaded, uncovered wagons in the street while they had lunch at a saloon. Reports were made to authorities, and when left unchecked, the drivers themselves were excoriated in the press.[41]

For the most part, though, the race of drivers and collectors is not mentioned in the city's annual reports, which seldom pointed to shortcomings of the drivers. The Magee-Flinn machine did not need to deflect blame because its position was so secure, as is illustrated by the failed legal case against the American Reduction Company brought by residents who lived on a bluff now polluted by odors from the plant. When a reform regime took over, it lacked the capacity to replace the American Reduction Company, so there was little subsequent criticism of the former regime's garbage collection. Complaints were directed toward householders who refused to separate their garbage properly.[42] But those householders were not singled out by race, class, or neighborhood.

Casting Blame in Charleston

Charleston resorted to race-based complaints in the early twentieth century, when there was swift back-and-forth capture of government by competing regimes. Charleston did not develop a new garbage collection program in the 1890s as other cities did. While other cities introduced new municipal garbage collection ordinances, Charleston kept delivering the services it had maintained for much of the nineteenth century. The mayor noted that the head of the scavenger division proudly reported that the garbage was picked up promptly and the force kept under "excellent discipline." The board of health warned that if garbage remained uncollected after midday, the driver would be blamed and possibly dismissed.[43] That discipline had always been a part of garbage collection in Charleston, where the drivers rang handbells and used red painted carts.[44]

Nevertheless, collection did not always go smoothly. When poor collection was noticed, officials looked to place the blame elsewhere. The street department blamed irregular garbage pickup on citizens who did not put out their garbage when they should. If there were problems, blame the households. The health department blamed Black residents for poor conditions and even deaths in their community. In 1887, Dr. Horlbeck ascribed the numbers to the loss of the paternal protections of slavery, attributing the cause "to the influx from a large surrounding population of colored people who came to the city for medical relief, and also to the recognized improvidence of the colored race. An examination of the decade from 1850 to 1860 will show a mortality among the white and colored about equal. Since the freedom of the negro race, the fostering care of the white being removed, we find an enormous increase in the death rate among the negro race."[45]

By the early twentieth century, Charleston was in the midst of the R. Goodwyn Rhett regime. Rhett was a descendant of Charleston's quasi-aristocracy who continued to dominate city government. The regime reorganized city services, placing the street department under the board of public works. It engaged in demonstrable discipline of city workers, outfitting them in "neat uniforms." Street sweepers followed the model of Waring's white duck uniforms and helmets. Drivers wore brown canvas suits with nickel buttons with raised lettering reading "D.S.C." (Department of Street Cleaning). Taking their new wages into account, Charleston required drivers to pay for the uniforms themselves. Enameled signs were displayed on collection carts, white letters on a blue background declaring "City" and the number of the cart. Once the garbage collectors were in uniform, their supervisor noted the "increased interest that they are taking in their work."[46]

Convict labor was not used in garbage collection, likely because of the proximity to the home, as garbage collectors retrieved and returned cans. The city used this labor only on the streets and in outer regions of the city. The collection of garbage nevertheless would have been secondarily affected by convict labor. Garbage collectors and drivers were included under a larger system of racial discipline. The source of city convict labor was the issuing of prison sentences for petty offenses. Given the need to muster suitable labor, the city now had an incentive to charge people for misdemeanors, including violations of garbage ordinances.[47] The internal system of uniforming, reprimanding, and threatening to fire workers likewise rested on the hierarchy of a racial order of white supervisors disciplining Black workers. The practice of convict leasing—renting prisoners to private industries—had been ended throughout the South, and this was hailed as reform, but prison labor continued, and was available for public works.[48]

While Charleston's use of convict labor did not extend to garbage collection, its racial disciplinary practices likely did. Hence the introduction of uniforms for street sweepers and garbage collectors in 1907 and 1908. Officials were quick to lapse into racial blame when faced with complaints about garbage collection. Their excuse was that "colored drivers" were "too careless" in handling the garbage. The health department recommended that carts be covered and that the drivers and collectors "be carefully watched."[49]

Despite the daily rounds of sanitary inspection, residents continued to generate a garbage can problem. Charleston's health department, for example, advised in 1903 that people should put their garbage in proper receptacles and leave them in a location accessible enough for the collector to pick up. A 1904 ordinance obliged householders to place their garbage in tightly covered receptacles. When residents failed to comply, the committee on garbage collections tried harder to regulate the behavior of householders, who were not using watertight receptacles despite "urgent appeals" from the committee and sanitary inspectors. The garbage collection department chided householders for failing to sort, putting tree trimmings in with the garbage when they could burn those, and mixing paper, excelsior, and straw in with the household garbage.[50]

The health department singled out residences in the Black neighborhoods, noting residents were "huddled together in lanes, rows and lots, as many as 100 to 200 in each such place, often in unfit and dilapidated structures, untenable except by Negroes of the poorest and lowest classes." These homes on poorly draining lowlands were subject to pools of collected water, inviting mosquitoes and disease. The health department recommended daily garbage collection from these neighborhoods.[51]

The racialized locus of blame diverted attention from the more systemic problems inherent in Charleston's garbage collection system. These complaints sound as though these problems were very frustrating to the Rhett regime, which had ostensibly engaged in institutional development with a board of public works, a new garbage ordinance, and a commitment of resources. They suggest that the residents—particularly Black residents—were holding back political development. Reading between the lines, we see signs that not all was right within local government agencies. The health department, for instance, pronounced that garbage collection was adequate, but the department wanted to return to daily collection, as had been done in the past. That was how things looked until John P. Grace took office as mayor in 1911. He hit the ground running, announcing his plan to bring the city forward, out of the Bourbonism that had held it back from development.[52] Grace won the election of 1911 narrowly, as a reform candidate unseating the old regime. His grasp on power was tenuous; he would lose reelection in 1915 but returned to office in 1919 for another term.

When Grace arrived in office in 1911, he abolished the board of public works, finding that it was too disconnected from democratic accountability, and he was careful in making appointments to the committee on streets, which was housed in the city council.[53] The Grace administration revealed how poorly garbage collection had been carried out by the prior administration. But the records of the prior administration lauded the efforts of city departments, casting the blame on noncompliant residents, especially the poor and people of color. Laying blame on people on the basis of race and class was a familiar resource for underperforming city departments to cover their own inadequacies.

Sanitarians hoped to improve city services, pointing out that the carts were not well adapted to their work; they were uncovered, spewing garbage through the streets, and were not made of watertight material. Though there were sanitary inspectors, there were not enough to enforce the laws requiring that garbage cans be placed on the streets only at designated times. Receptacles were scrutinized without criticism of householders. Not that blame on residents was entirely absent. While the Charleston street department recorded an increase in the number of loads of garbage collected, health officials complained that garbage collection was "crude and unsanitary." This "antiquated and inefficient" system with uncovered carts left "unsanitary mass upon the streets, to be scattered by negroes and dogs and cats."[54]

Mayor Grace unloaded his frustration in his 1914 annual address. He charged that the old regime had sabotaged his administration from its first days, as it "attempted the disruption of our forces by sowing discord and temptation amongst them," leaving him to contend with "many silly jealousies and rebellions."[55] It was within this troubled space that the civic clubs managed to find entry. By 1914 the Charleston Civic Club, like its counterpart in Pittsburgh, and its Junior Civic League went to work on providing garbage receptacles in public places. They gained some access to city officials and introduced a petition requesting appropriations for necessary painting and repair of cans.[56]

Grace lost the 1915 election to Tristram Hyde, whose administration reverted to the old ways of doing things. Health officials once again sought a "guilty party" for a messy garbage situation, reporting in 1916, "We have tried to educate people to the use of the metal cans and have obtained some results, the colored people giving us the most trouble in this respect." The next year the street department worked with the Civic Club on a Clean-Up Campaign, with satisfactory results: carts were now covered. The remaining problem was the people. They had been educated on the use of metal cans and still they did not use them. Their garbage spilled out into their yards, and "there are places in the heart of the city so congested with colored [sic] people that it is almost impossible to find the guilty party and it keeps our small force very busy following them up."[57]

Like the Rhett administration, the Hyde administration may not have picked up garbage cleanly, but it certainly advanced institutional development. Charleston had historically used garbage for fill in lowlands, or dumped it beyond the city limits, but as the city expanded, these methods were becoming untenable. Back in the 1880s, Dr. Horlbeck had urged the city to switch to incineration. The Hyde administration put out bids and finally built the "Destructor." Initially, results were favorable. The only fuel it needed to operate was the refuse it consumed, and the chimney did not pollute the air. On October 29, 1919, however, the Destructor was closed down for repairs until September 1920. Collectors once again dumped garbage on the salt marshes. When the incinerator was back in operation, it was able to consume only 15 percent of the garbage.[58] By 1921, the Destructor ceased operating.

San Francisco Scavengers Maintain Control

San Francisco used an entirely different mode of municipal garbage collection. As San Francisco developed in the 1850s, its hilly topography presented issues for disposing of garbage, wastewater, and human waste. With privy vaults and surface sewers overflowing, the city needed to rely on private capital to fund sewer construction.[59] In 1856 a vigilante group, the Vigilance Committee, was revived with the aim of reining in government corruption. The organization took over city government and took credit for slashing expenditures and cutting taxes.[60] This regime attended to water delivery and sewers but not successfully, as the sewerage both smelled and harbored disease.[61] The city never took over garbage pickup, and San Francisco was, technically, a city with no municipal collection. The garbage did get collected, however, by scavengers who made one-on-one arrangements with individual households to cart off their waste. The scavenging business was run by an organized group of Italian immigrants. The scavengers did not work for the city, nor were they contracted by the city, but the city relied on their services.[62] Leaving trash collection to private contracts, the City of San Francisco did not invoke race to meet its objectives. Scavengers, however, did.

In the 1870s and 1880s the city had to contend with scavengers who furtively deposited night soil on empty lots or even in the streets and sidewalks.[63] In 1885 the health officer's annual report expressed concerns about diphtheria, overflowing sewers, and the "disgraceful mode of disposing of the garbage of the city" by dumping, and noted how cheap it would be to dump garbage out at sea.[64] One member of the Board of Supervisors got tough on garbage, proposing that householders collect their ashes once a week in metal containers. Anyone caught violating the ordinance would be fined. The measure was opposed for its disparate impact on poor families.[65] The following year the Health and Police Committee of the Board of Supervisors turned on the scavengers, requiring them to load their collected garbage onto barges instead of dumping their loads on North Beach lots. Since scavengers were freelancers, basically men with a horse and cart, they lacked the capacity to comply. They promised to build their own boats but did not follow through.[66] In 1887 the city again expressed a wish to stop dumping, but dumping was expedient for scavengers, and it was cheap as well. Local officials wanted to have garbage hauled away on barges, but scavengers did not own them. A barge operator would charge the scavengers, who would then pass on the cost to their householders. The scavengers balked and refused to comply. The street department urged that "something had to be done or the scavengers would soon cause the town to become as fragrant as the city of Cologne."[67] City officials gave in to the scavengers' union.

In 1898–1899 the Board of Supervisors tried to make good on its contract with the Sanitary Reduction Works for incineration of garbage. The terms of the contract dictated that the incineration plant would yield revenue for the city without increasing the cost to householders.

When the cost was passed along to householders after all, the scavengers objected to discrepancies in measurement of the garbage they brought to the works, leading to conflict between the scavengers and the sanitary works. When local officials decided to award a contract to the company for exclusive responsibility for disposal, the city tried "a new scheme to circumvent the scavengers."[68] The scavengers responded by working with the rival California Reduction Company of Colorado, which was incorporated in Colorado in order to be able to take legal disputes into federal court. That company offered to take the scavengers' garbage, load it onto barges, and dump it at sea. When the California Reduction Company filed suit to challenge the Sanitary Reduction Works' exclusive right to receive garbage, the judge ruled in favor of the Sanitary Reduction Works.[69]

The city could not wrest control from the scavengers. It could only try to check them. Scavengers were known to dump garbage in vacant lots, using "unlined, filthy, ill-smelling, leaking wagons." Sanitary inspectors were put to work to catch them in the middle of the night and gain convictions. The chief sanitary inspector noted that this work was done in the dirtiest parts of the city, namely, Chinatown. This was, he said, "disagreeable, and I emphatically add DANGEROUS, work necessary in that district, the most degraded, filthy, immoral and unsanitary location within the boundaries of the United States of America."[70]

Scavengers were instrumental in the cleanup after the 1906 earthquake, when the need for sanitation was critical. While 463 deaths were reported as a result of the earthquake, another 547 deaths occurred soon after from typhoid fever.[71] Disinfecting crews were sent out to spray, dig up latrines, and collect debris. "Housewives who still had a house" poisoned rats in the home while soldiers bayoneted them.[72] Residents were instructed to leave out their garbage barrels at the curb. For thirty days, two hundred teams and 225 men were enlisted, at no cost to householders, to remove seven hundred loads of garbage daily.[73] When the incinerator reached capacity, garbage was loaded onto barges and dumped at sea.[74]

Competing scavenger associations fought for control over garbage collection. The scavenging business ended up becoming dominated by immigrants from northern Italy.[75] The city tried again to break up the scavengers' monopoly in 1919. A special committee on garbage disposal studied the rates charged for garbage collection. It conducted a time study measuring how long it took for a driver to dismount from his wagon, walk to the curb, walk to the can, and so on. Although scavengers competed with one another, the city's interest in setting rates compelled the scavengers to close ranks. They formed their first cooperative in 1920—the Sunset Scavenger Company of San Francisco. The Scavengers Protective Association followed in 1921.[76] They were able to work with the city, obtaining the exclusive license for refuse collection.

But the "garbage wars" flared. In 1929 San Francisco tried to crack down on scavengers overcharging by requiring them to print their rates on the back of receipts. Faced with a loss of their collective hold on garbage collection to municipal collection, the scavengers reached out to housewives, letting them know how reform would affect them. A flyer from the City Sanitation Guild (one of the competing scavenger associations) was titled "Housecleaning" (figure 6.1).

It announced that the scavengers had "cleaned house," but this was not the "feather duster variety." This housecleaning had removed political members and installed new officers "whose standards are those of public service." It was no accident that the guild's words are suggestive of municipal housekeeping. The flyer continues: "Housewives know that once the spots are removed the house is bright and shining again. The City Sanitation Guild has profited by their example and will continue its policy of giving the best possible sanitary service to the people of San Francisco."[77] The Sanitation Guild went further in another flyer (figure 6.2), directly appealing to women and letting them know that the guild needed their support:

> Every woman who manages a home knows what the breakdown of a garbage system, one which has been developed over a period of fifty years of service, would mean to the health of the community. San Francisco's scavengers are confronted with the possibility of such an emergency if permit traffickers, posing as garbage collection agencies, are allowed to operate in San Francisco. In the long run they force the City Sanitation Guild to buy them out—this occurred in the past—and the HOUSEHOLDER must then bear part of the burden of this expense.[78]

Another flyer in May 1929 played upon their customers' deeper fears. Should the city impose this competition on the scavengers, it would open the door to the "great danger to the San Francisco housewife lurking in the proposed plan to make garbage collection a huge public project with a new force, under public employ, operating the collection system. *It would mean that every housewife would be at the mercy of marauders and prowlers*" (figure 6.3).[79]

HOUSECLEANING

The Scavengers of San Francisco have cleaned house. It has not been the feather duster variety of spring cleaning ✎✎✎ thorough scouring and cleansing have been employed so that officers who were playing politics have been removed. In their places have been installed new officers whose standards are those of public service.

Service has been the motto of San Francisco Scavengers for a period of fifty years. The record of this body has been marred only by the actions of publicity seeking individuals and extortionists.

San Francisco Scavengers have therefore instituted their house cleaning policy. They have resolved to take the public, whom they have served so long, into their confidence.

Housewives know that once the spots are removed the house is bright and shining again. The City Sanitation Guild has profited by their example and will continue its policy of giving the best possible sanitary service to the people of San Francisco.

City Sanitation Guild
[Scavengers Protective Union]

PATRONIZE OUR ADVERTISERS

FIGURE 6.1. The garbage wars in San Francisco

Source: City Sanitation Guild, "Housecleaning," flyer dated February 1929, Garbage—City Sanitation Guild, 1929, San Francisco Ephemera Collection, San Francisco History Center, San Francisco Public Library.

March 1929

WE   C.S.G.   YOUR
NEED   HELP

City Sanitation Guild
Scavengers' Protective Union

San Francisco scavengers have done all that is within their power to find a solution for the garbage problem which confronts the city. They have offered to print a full schedule of rates on the receipts which are given to the householder (even though these are unscientific)

BUT

Their efforts will be lost unless they have the support of the WOMEN of San Francisco.

Every woman who manages a home knows what the breakdown of a garbage collection system, one which has been developed over a period of fifty years of service, would mean to the health of the community. San Francisco's scavengers are confronted with the possibility of such an emergency if permit traffickers, posing as garbage collection agencies, are allowed to oper-ate in San Francisco. In the long run they force the City Sanitation Guild to buy them out—this occurred in the past—and the HOUSEHOLDER must then bear part of the burden of this expense.

Your scavengers are unwilling to subject you to such expense. They are asking the women of the city to band together and express themselves, through their various organizations, so that a happy solution may be real-ized.

City Sanitation Guild
(Fifty years of service to San Francisco)

FIGURE 6.2. Scavengers appealed to householders' security and cost concerns
Source: City Sanitation Guild, "We Need Your Help," flyer dated March 1929, Garbage—City Sanitation Guild, 1929, San Francisco Ephemera Collection, San Francisco History Center, San Francisco Public Library.

A PLAIN TALK
TO
HOUSEWIVES
ABOUT THE
GARBAGE
SITUATION

There is **great danger** to the San Francisco housewife lurking in the proposed plan to make garbage collection a huge public project with a new force, under public employ, operating the collection system.

It would mean that every housewife would be at the mercy of marauders and prowlers.

Unknown men and untried workers, a floating population of itinerant help, would be at liberty to wander about your property in search of "garbage." Cities in which such a plan has been tried have suffered waves of petty thievery and attempted violation of women.

> There is only one band of workers in San Francisco at the present time which is legally authorized to collect your garbage.

These men are tied together by the City Sanitation Guild.

They are responsible business men. Just like your grocer, they have an investment to protect.

> They are a stable part of San Francisco's population. Each man owns his own equipment, each is maintaining a home and the education of children.

It is for such workers that we solicit the support of the housewife, as against a roving band of vagrants which threatens to invade the city and place garbage collection—upon which the very health of the community depends—in the hands of politicians.

HELP US TO HELP YOU

City Sanitation Guild

FIGURE 6.3. Scavengers exploited the fears of housewives

Source: City Sanitation Guild, "A Plain Talk to Housewives about the Garbage Situation," flyer dated May 1929, Garbage—City Sanitation Guild, 1929, San Francisco Ephemera Collection, San Francisco History Center, San Francisco Public Library.

WHAT YOU
OWE
YOUR
SCAVENGER

As householders of San Francisco, the women of the city have played and have still to play an outstanding role in solution of the garbage situation.

You owe to us your allegiance and support because we have for fifty years given ours to you.

It is the women of San Francisco, more especially the housewives and mothers who must shoulder the responsibility of preserving for the city a scavenger service which has ALREADY PROVEN ITS WORTH—for—

THERE IS STILL MUCH TO BE DONE

if the carefully worked out garbage collection system—a system which has pioneered in the field of sanitation in this city—is to continue to operate unimpaired.

Unless you support the scavengers who now serve you and whose fathers and even fathers' fathers served past generations of San Franciscans, they cannot withstand the ruthless competition which threatens to paralyze and break down an organization which has stood the test of time.

Therefore we ask you to

STICK TO YOUR OWN GARBAGE MAN

He is trained in his job and eager to serve you.

City Sanitation Guild

FIGURE 6.4. Scavengers asked for loyalty

Source: City Sanitation Guild, "What You Owe Your Scavenger," undated flyer, Garbage—City Sanitation Guild, 1929, San Francisco Ephemera Collection, San Francisco History Center, San Francisco Public Library.

In their appeal to housewives, the scavengers, embroiled in a potential takeover by the city, played upon women's fears of having their homes approached by unknown men. In response to those fears, the scavengers encouraged women to "STICK TO YOUR OWN GARBAGE MAN" (figure 6.4).[80] The fear-inducing messages that San Francisco scavengers directed toward housewives illustrate the intersection between race and gender in garbage collection and the various ways it could be manipulated by different parties. The scavengers once again successfully resisted city takeover and continued to operate San Francisco's garbage collection.

Racial Hierarchy as Resource

Racial disparities played an important part in nineteenth-century cities: determining where people lived, what they did for a living, and what rights and privileges they had access to. Racial hierarchy was marshaled by governing authorities to discredit scavengers and push for policy change, and it was used to place or divert blame for poor garbage collection and disposal. What is remarkable is how little race is explicitly mentioned while remaining available as a resource, invoked particularly when regimes felt vulnerable or attacked. Racial hierarchy remained a useful tool of governance, whether for sanitarians to seize control of nascent collection programs or for established regimes to deflect blame from their lack of will or ability to carry out garbage collection successfully.

CHAPTER 7

The Politics of Garbage Collection
Lessons Learned

How did municipal governments across the United States deal with mounting trash in the late nineteenth century? They often ignored technical experts—sanitarians, boards of health, sanitary and other engineers—and civic associations that pointed out the problem in the first place. Instead, when corrupt actors decided to act and what solutions they chose depended on how they could benefit politically and financially. But once policies were put in place, non-democratic political leaders still faced challenges with residents' noncompliance and political blame when programs came up short. Political officials drew on gender and racial hierarchies to address implementation and its failures. More than a century later, Americans still use the sanitary infrastructure and strategies these actors put in place, but the history of trash and the messy politics that surrounds it has largely been lost. Americans take out their trash on the appointed day, at the required time, in the specified container, and it's all seen as a matter of individual habit without any sign of coercive political power or shaping of civic identity.

Garbage collection is both ubiquitous and ordinary, the very place where governing often happens but is largely invisible. It's only when collection fails—when the bags pile up and spill out—that people stop and take notice. When we pause to consider what went into the creation and maintenance of garbage collection and disposal programs, we may be encouraged to rethink what is political in the first place. Politics does not happen only in faraway national or state capitals. It is right in front of us in commonplace practices. We can see everyday politics when we expand our scope to think about how governments operate, what resources they rely on, what development really means, how people are affected in what they do in their own homes and in their bodies, and how they are perceived by others, depending on how they are enlisted by municipal governments in carrying out their goals.

The history of municipal trash collection and disposal offers three lessons for scholars of politics, power, and development: (1) how potent policy problems and heated political disagreements come to be seen as nonpolitical or even mundane over time; (2) how governments use both formal resources (agencies, expertise) and informal, unsavory resources (corruption, gender and racial hierarchies) in public policy programs; and (3) how political development includes not only the steady march of liberal values but also the long-term incorporation of undemocratic elements. Our work draws on and builds upon a broader literature that examines politics at the "margins," understood both in the marginal topic of garbage collection and in the unexpected resources brought to bear on politics and development.[1] In this chapter we discuss lessons from our analysis of nineteenth-century trash collection and how they fit into what we already know about government capacity to provide services, the resources governments draw on, and the long-term implications of both for political development.

State Capacity

Even where municipalities were able to pass garbage ordinances, problems arose when they tried to pick up trash, cleanly, throughout the city, week after week. These local governments ran into questions of capacity. State officials may choose not to address problems that they have no ability to solve. They may create public programs that they have no ability to implement.[2] What state officials *want* to do and what they are *able* to do may be very different. State capacity, according to Theda Skocpol, is the ability "to implement official goals, especially over the actual or potential opposition of powerful social groups or in the face of recalcitrant socioeconomic circumstances."[3] State capacity is a concept shared by the fields of public administration, sociology, and political science (including comparative politics and American political development). As a result, there is wide variation in the definition and operationalization of the concept. In a review of the broader literature, Luciana Cingolani summarizes state capacity as "the state's ability to credibly implement *official goals against potential resistance, by means of taxation, the bureaucracy, the military, the legislature, and the courts, and throughout its entire domains.*"[4]

State Capacity as Administrative Capacity

American political development (APD) scholars have largely examined state capacity in terms of administrative capacity and transformation of the federal government. Stephen Skowronek shows how the United States transformed from a nineteenth-century patchwork state of "courts and parties" to a twentieth-century state built on administrative power.[5] At the heart of administrative capacity is the federal bureaucracies' technocratic capabilities. Daniel Carpenter, in a study of the United States Post Office, Department of Agriculture, and Department of the Interior, describes how bureaucratic federal agencies beginning in the twentieth century began to "forecast, plan, gather, and analyze intricate statistical information, and they execute complex programs." They are able to accomplish as much as they do by relying on analytic (informational) and programmatic (planning) capacities.[6]

Yet government-provided services do not always equate with administrative capacity the way that scholars like Skowronek and Carpenter have shown. Although local governments created and oversaw garbage collection through municipal agencies—departments of public works, health, or streets—just as often local governments contracted out municipal collection or opted for none at all, deferring to private scavengers.[7] And contrary to presumptions that government-provided services are indicative of capacity, some of those alternative methods did a much better job at picking up garbage citywide. Therefore our study of the development of garbage collection is about not only formal administrative capacity but also available resources that governments leverage whether formal or informal, public or private, democratic or not.

The Nineteenth-Century Hidden State

APD scholars have demonstrated that there has long been a state, and state capacity, often hiding the reach of its activities. In a study of the Indian Vaccination Act, Ruth Bloch Rubin finds pockets of real capacity

169 THE POLITICS OF GARBAGE COLLECTION

in the antebellum federal state, although in places we might not expect. She illustrates the important role the War Department played in administering vaccinations to Native populations. William Adler, too, shows the importance of the War Department in antebellum America by identifying the role of the Army Corps of Topographical Engineers in economic development policies. Andrew Kelly looks at the development of top-notch US scientific expertise during the nineteenth century, a time often characterized as having a weak administrative state. Brian Balogh has suggested that the federal government "governed *less visibly*" in the nineteenth century. The federal government used its authority as coordinator of state and local activities at the literal margins of land, or in judicial rulings that allowed for growth of large corporations and national markets. Balogh sees state building even though there was not much federal administrative state to speak of.[8]

For all of the emphasis on national agencies, localities have been the place to exercise police powers, whereby residents' lives were shaped to further "the well-regulated society."[9] State and local governments were vitally important in the late nineteenth and early twentieth centuries, particularly for public health.[10] State-level agency modernization was sometimes purposely overlooked, even at the time when Progressive Era reformers obfuscated such development to distance their programs from local-level corruption.[11] The federal government was not so much absent as it was dependent on lower levels of government. The federal government relied on state and local governments as resources. These administrative agencies and lower-level structures of the hidden state are all formal, public institutions.

Private actors, too, are involved in the development of state capacity. Colin Moore shows how American imperialism developed through public-private partnerships with investment bankers, which were created and supported "through a variety of cross-cutting professional networks."[12] Andrew Kelly details how US scientific capacity evolved when public agencies (the United States Coast Guard Survey and the United States Geological Survey) partnered with private actors and institutions, which were integral "in the development, administration, and direct undertaking of federal scientific initiatives." The public-private relationships define scientific initiatives in the United States to this day.[13] Non-state actors have been present in the development of administrative agencies and policies, whether maternalist reformers laying the groundwork for the welfare state or agencies reaching out to networks.[14] Daniel Carpenter suggests looking not just at the legislation but at the operation of bureaucracies and the relationships that bureaucrats forged.[15] Scholars continue to identify the hidden state in the welfare state by pointing to mechanisms such as taxing as part and parcel of the submerged state.[16]

Incorporating public-private relations takes us to the boundaries, where actors' behavior might not even make them likely candidates for public-private partnerships. A civic organization, and a women's one at that, such as the Immigrants' Protective League, can be overlooked, as can its efforts to have the state take on new tasks and identify new public missions. Carol Nackenoff reminds us, "Public-private collaborations are important components of state building, and the impact of non-state actors is larger than what is commonly claimed."[17] Oftentimes these collaborations are at the state and local level, but they can diffuse far beyond that. State and local actors can create policies, programs, or strategies that are then adopted by other state and local or federal actors.[18]

Within these margins, too, are power dynamics, and when the state draws from the margins, it taps into conditions of inequality. Sometimes public agencies collaborating with private actors incorporate the biases of these private actors.[19] Historians who have looked at the boundaries of the state note as well the agency possible in that relationship, with those actors on the boundaries reframing state power and sending it back to the state.[20] Boundary organizations might be voluntary associations that are handed power by the state, so that a dog-catching organization implemented public health measures.[21] Or they might be participatory organizations like 4-H, which was designed to acculturate rural youth into agriculture but could also be used to shape their sexuality and their acceptance of federal expertise.[22] Those on the margins are not ignorant of the state; they may be more aware of the operation of hidden state power because they feel its operation so acutely.[23]

Carl Zimring has recovered those scavengers who were further marginalized as garbage collection modernized.[24] We build upon such recovery to identify private actors and politically constructed social relations as resources in garbage collection in the generation of municipal ordinances, in contracts, as targets of political blame, and as model citizens in the deployment of infrastructural power. By acknowledging these relations between formal state institutions and private actors as resources of governments, we can categorize what we have found as both extractive capacity in the government's ability to gather resources as well as the capacity to induce compliance.[25]

Corruption as Extractive Capacity

The country's public health changed dramatically as a result of sanitary improvements, including garbage collection. But these positive outcomes were not the result of the federal government's administrative capacity or even of the technical expertise of a growing cadre of national professional organizations. Instead, the transformation of American public health resulted from the aggregation of municipal decision making, often by corrupt governments across the country.

Although the end of the spoils system might be expected to erect administrative capacity or create public agents who must keep a line between public and private interests, our own study of nineteenth-century municipal trash collection shows that corruption can promote state capacity for weak governments. Certainly, corruption can lead to failure, as illustrated in St. Louis, which developed capacity to collect and dispose of trash only to lose it in the fight against corruption, and New Orleans, which failed to develop any modern, sanitary methods of trash collection and disposal at all. But corruption also provided a bridge to capacity in Charleston, which developed an early system of trash collection by leasing slave labor, and Pittsburgh, where the machine contracted out to a machine-run business. In both Charleston and Pittsburgh, political actors were able to leverage private resources *because* they had corrupt ties. The trash collection and disposal programs they created lasted. Reform regimes adapted the existing infrastructure and continued to provide services for decades.

Whether well-meaning agencies are seeking to expand their authority or merely accomplish policy goals they were tasked with, studies of state capacity assume organizations must be free from corruption. Both Skowronek and Carpenter detail how building the administrative capacity of the twentieth-century American state was predicated on overcoming the corruption of the spoils system, characterized by rotation in and out of government for party loyalists after an election and assessments paid by employees to the party.[26] Corruption was the target of progressive reform. Carpenter recognizes that bureaucratic outreach to networks in business and society was necessary to build administrative capacity, measuring capacity as an agency's autonomy from inefficiency or corruption because his standard is agency autonomy and the legitimacy that comes from reputation.[27] Agencies that do not develop reputations lead to policy failures. Our focus on capacity as the ability to garner resources allows us to see the work that corruption can do when a government is unable to do that work itself. That is, corruption is not a deviation from what the state has decided to do but a means to accomplish what the state has committed to.

Because a constantly shifting workforce chosen for its party loyalty is not necessarily capable, let alone concerned with a well-functioning state, it is no surprise that Carpenter defines capacity as "the collective talent of bureaucracies to perform with competence and without corruption and malfeasance."[28] Moore, in his study of imperial policy, notes that "partnership requires that bureaucrats remain independent from private interests and retain their own unique preferences, which they then pursue through these partnerships."[29] Yet as we show, corruption can be an available resource too. Corruption motivated regimes to take on public works, including a willingness to invest in innovative technologies. Our point is not to bring corruption back into vogue but to acknowledge that corruption can be a feature of political development and not something left behind.

Capacity to Induce Compliance

The reliance on private actors—whether elevating middle-class white women or denigrating poor neighborhoods—was a product of governments' use of resources to accomplish their objectives. In the process, they produced differentiated citizenship. Beyond administrative capacity for implementing federal public programs, scholars have also looked at the ways that state capacity varies across the United States and for particular populations within it. They show that national policy objectives can look very different when they are implemented by lower levels of government.[30] State capacity is built through differential policy enforcement, and state power is wielded in ways that can produce differentiated citizenship.[31]

The hidden uses of power by a government can rely on and reshape political identities based on race, class, and gender. APD has recognized the endurance of ascriptive status and the structural survival of such ascriptions in racial orders.[32] Such orders can be enlisted as capacity to mark populations and further differentiate them. Chloe Thurston points out that in the era of the weak federal state, military power was used within US borders to "control the settlement, movement, and life chances of marginalized groups." Not only was this power hidden from public view, but also it was differentiated, with groups experiencing different effects of coercive state power.[33] Thurston's analysis highlights how the hidden state is also constructing identity, which is why a study of racial politics looks not just to different sites but to different time points. The moments that are considered transformational in APD may not be the moments that capture these institutional uses of racial reliance and construction.[34]

Our study of the generation of capacity to collect garbage reveals racial politics, where racial identity is not a given but is constructed through these processes.[35] In relying on existing racial hierarchy, the state is likewise engaged in racial formation.[36] Furthermore, these constructed identities are incorporated into institutions because they are resources of governance. How does this happen? Let's take the habit of putting out a garbage can. People have internalized the rules of putting out a garbage can: what can to use, how to fill it, where to put it, and when. Those habits creep back into the home: sorting recycling in the kitchen, having appropriate kitchen cans to facilitate any sorting, and hauling on garbage day. All of these actions seem like personal habits and consumer choice. Yet when we contrast them to New Orleans residents' outrage over garbage boxes or Pittsburgh's yards of overflowing, uncovered cans and boxes, we recognize that household habits can be a feat of governance. The generation of those habits rested on tools of gender and racial hierarchies, reproducing privilege, and then obscuring it to make the political look personal.

Development at the Margins

Scholars commonly agree that political development is the durable shift in governing authority,[37] but they diverge from there. One body of research examines the growth and progress of the US national government, focusing on prominent agencies and larger-than-life actors. A second body of research examines development in America at the margins, at lower levels of government, recovering actors and agencies largely forgotten by history. Rather than chronicling the advance of robust federal agencies with familiar names, Carol Nackenoff and Julie Novkov describe the "muddled mix of local, state, national, public, and private interests in policies, and a range of actors seeking leverage on policy in any ways and through any arenas where they could find it."[38]

Our study of nineteenth-century municipal trash collection builds on the latter tradition. We argue, however, that while local garbage ordinances may seem to be marginal, the ubiquity of trash collection programs and the effect they have had on American public health is a story at the heart of political development, the shift from private action to public responsibility.

Indeed, the infrastructure of people's lives is felt most closely in state and local—not national—policy domains. A growing body of scholarship unearths the mechanisms of local governing that provide these services, unevenly, and shape citizenship. Jessica Trounstine uses sewerage overflows to demonstrate how segregated cities produce the material conditions of different populations and then attitudes about them. Her study shows that when local governments mete out basic services differently to different populations, they disparately affect residents' daily lives. Furthermore, these unequal provisions of public services generate perceptions of bustling, thriving communities versus blighted, struggling communities. Colin Gordon points to the use of annexation, zoning, and redevelopment as spatial tools to provide housing and services unevenly along racial lines, fostering a fragmented local citizenship.[39] City governments, strapped for cash, have exploited this differentiated citizenship that they created to discipline, excessively fine, and commit violence against their own residents.[40] We find that the connection between public services, racial hierarchy, and sanitation outcomes gets buried, naturalized, and personalized.[41] Recovering that connection reveals the governing authority that is part of this regime, not an exception to it. The identification of slackers who generated the garbage can problem, the pronouncement of blight to justify redevelopment programs that categorized residents by race and class against a white suburban ideal and used government authority to displace them, share a deployment of racial politics by municipal authorities to achieve public purposes.[42] Although these are features of local political development, because they happen systematically across the country, they constitute American political development.

Kimberley Johnson too shows the importance of development through state and local actors. She suggests that the idea that development is contrary to corruption, cronyism, and nascent resources is itself a product of Progressive Era reformers. State boards and commissions did the work of governing into the twentieth century. They confronted fears of the state while administering policies and modernizing administration.[43] Governing happens in the states. The institutional arrangements, networks, and rules of state-level administration shaped the introduction and modernization of policies even before they could be taken up in the New Deal order.[44]

Johnson reminds us that when we pay attention to local and state operations, we see racial politics develop into racial regimes. Rather than view the South as a fixed region in "suspended animation," she finds that struggle occurred, and again and again "it was stateways that gave the color line the force of law."[45] At the state and local levels, policies from social services to policing produce differentiated citizenship.[46] National policies may be implemented locally, and then that becomes the site of citizens' experience of state authority.

As scholars "bring the city back in" to the study of governance and political development, they identify features of governance particular to urban areas, such as the role of capitalism in conceptions of growth, or the role of coalitions in civil society enacting policies.[47] While these features might be present at the federal level too, their operation will look different in the environment and imperatives of the local. Local government and local actors do not operate in a vacuum. Recognizing the role of national institutional structures, such as bond markets in infrastructure investments or competition among cities, Neil Brenner suggests incorporating national state power to view the operation of cities.[48] In pursuing the enactment of ideas in policies or the binding up of ideas in institutions, scholars of urban politics and urban development demonstrate that local conditions and local implementation shape national policies, whether formal civil rights laws or ideologies such as neoliberalism.[49] As shown by the work of Trounstine and our work here, local actors create national-level outcomes when similar policies and practices diffuse across the country.

Nackenoff and Novkov, too, see not just pre-national policy but the power relations at these levels. These are places where people at the margins may have exercised agency and devised new visions for what a government should and could do. The inclusion of these social relations into various levels of government institutions makes for a muddled state, but this is the feature, not the bug. The American state works like a Rube Goldberg apparatus.[50] The private actors engaged in relations with the state, then, affect development through their visions, resources, struggles for equity, their dependence on others, and so on.[51] Jamila Michener, Mallory SoRelle, and Chloe Thurston suggest fundamentally changing the starting point and beginning inquiry from the perspective of those at the margins rather than approach them top-down. Thus we can identify those who may have organized for these policies in the first place, who feel the power of the state when policies are enacted, and who may engage in behavior that seems surprising but makes sense as adaptation to policies. When we do, we see government authority operating "in the lives of ordinary people."[52] We did not set out to look at ordinary residents as Michener, SoRelle, and Thurston suggest, but we caught glimpses of resistance and shaping of policies at the margins. Using formal documents of cities, we explained how municipalities assembled resources to meet the garbage problem. It was reading between the lines of official documents, or wondering why expected patterns of development did not fall into place, that led us to see how governments reach out to the edges to meet the imperatives of governance.

Conclusion
Everyday Politics in Practice

A wave of municipal garbage ordinances spread across the United States in the 1890s. But it's not as if there had been nothing done with garbage before then. City residents always had some way to deal with garbage. People may have sold it to a farmer who came through their streets to feed it to his hogs, or paid a scavenger to haul it away. Others buried it in their privies or threw it into a burn pile. As populations increased, these methods didn't work anymore. In the 1890s political development meant that garbage collection and disposal shifted from private to public authority. Municipalities generated political will and administrative capacity for collecting garbage at the local level. While development in garbage collection and disposal did not happen at the federal level, it did happen nationwide. The aggregation of discrete municipal actions—at roughly the same time in cities across the country—developed the nation's sanitary infrastructure. City residents experienced the shift in sanitary practices as part of their everyday life.

The garbage problem was a challenge for growing cities, and its solution was available in the progressive time in which it occurred. There was no shortage of expert knowledge ready to be directed toward the public good. Sanitarians offered information and advice on safely collecting and disposing of garbage based on the best available evidence.

Engineers designed devices to collect and dispose of trash efficiently. Civic reformers stepped forward to promote these advances. Yet city governments often sidelined expert advice. They responded with a wave of municipal ordinances but were faced with practical challenges: Would they contract out or develop their own departments of public works? They might need to share horses and carts with departments of streets. They needed to hire, train, and manage drivers and collectors. They needed to figure out whether the waste would be carted to an outlying dump or the nearest body of water, or if they ought to develop some kind of incinerator or even a reduction plant. Garbage collection and disposal required technical expertise and skills, political will, ability, compliance, and political cover.

Cities had their own "prior ground": Some already had horses and carts from other departments. Some had means of exploiting labor. Some had sanitarians housed in the board of health. Some had top-notch organized systems of scavengers who just needed a license from the city, so the city didn't have to do much at all. As each city looked around for its own collection of resources and the ability to harness more, it did so within the context of its own political regime. In the 1890s there was a good chance that the regime was corrupt. Corruption is a broad term that can include a variety of forms, whether political machine, aristocracy, oligarchy, or ring, which in turn used various mechanisms (rigged contracts, nepotism, boodle, patronage) to achieve their objectives. Yet when we focus on the monopolization of political power (rather than on a dichotomy of corrupt versus reform), just about any regime can be seen as using this new municipal program to its own benefit.[1] When cities adopted a progressive project, the imperative to carry it out sent them casting about for the resources available to this particular regime in this particular city.

This book has examined the resources that cities used to develop trash collection and disposal programs. Some governments found their particular form of corruption to be a resource. A machine boss could throw the contract to a machine-connected business, and the machine would profit. But it also set up a collection and disposal method that even reformers could not replace and had to rely on until they were able to marshal their own resources decades later. A city with a history of exploiting human beings in slave labor could modernize that subjection of Black citizens while prolonging the dominance of the leading white families. Corruption was not necessarily an impediment to governments' achieving their objectives; it was a resource. Whether it was an effective resource or not depended, we find, on the integration of the regime within the government.

The development of administrative capacity did not modernize citizenship when resources relied on hierarchies that crossed race, class, and gender lines. We found that when city governments were faced with a lack of capacity and looked around for resources, they discovered that these categories and inequalities were available. Gender hierarchy was enlisted to help solve the garbage can problem, as some women were invited to offer themselves as model residents and base their civic identity on following rules and encouraging others who were different from them to be more like them. Poor neighborhoods and people of color were mentioned in annual reports in years when governments delivered shoddy service. Casting blame on badly behaving neighborhoods could help an administration deflect blame.

Such mechanisms of governance were folded into garbage collection and disposal programs. When Americans stopped noticing garbage as a political problem or an exercise of political power, those mechanisms were hidden as well. Garbage collection is among the most basic and seemingly innocuous operations of governing. Yet the history of the programs suggests a very different, far more contentious, confrontational, and complicated story. When we look at the development of garbage collection and disposal programs, we can identify the basic tools of governments. In this book we show which resources governments use and for what purposes they use them. It is not a story of political actors, agencies, and experts alone. Instead, it involves undemocratic and even unsavory resources that they enlist in the course of political development.

The wave of garbage collection and disposal ordinances passed in the 1890s was not just a response to growing awareness of the need for better sanitation and suggestions for how to respond to it. Merely because there is a problem does not mean that anything will be done about it.[2] Cities needed political will from officials in charge. Corruption generated that political will when regime leaders realized they could benefit financially or politically. Cities also needed the administrative capacity to collect and dispose of trash, whether to administer the program through a department of streets, public works, or health, or contract it out. Municipal executive branches relied on both formal and informal resources, including corrupt relationships and arrangements. Once programs were created, though, implementation proved to be a challenge. City leaders struggled with resident compliance and with political blame for slipshod services. They similarly relied on formal and informal resources. Governance involved politics—excluding political opposition from the execution of collection, privately benefiting from public allocations, reproducing social hierarchies as efforts to achieve compliance focused a lens on good citizens and those to be derided for their unclean homes—including dirty politics of politically corrupt regimes. Governments used power to shape political identity.

Whatever obstacles garbage collection programs encountered in their early development, they eventually settled into place, passing from regime to regime without disruption. In fact, garbage collection today looks much as it did back in the late nineteenth century. Development started at the local level and stayed there. It drew its resources from the

FIGURE 8.1. Pittsburgh garbage collectors, circa 1960s
Source: Bud Harris Collection, Archives Service Center, University of Pittsburgh.

local level. But the effect is not simply local. Cities across the country engaged in similar—albeit not uniform—behavior, transforming American sanitation and health outcomes. Municipal residents gained new services at the same time they were being shaped by their relation with government, whether they knew it or not at the time and whether or not they do today.

It is in these ordinary activities that people engage in politics.[3] In the local and in the mundane (like garbage collection), they engage with government, comply with it, gauge their behavior with and against their neighbors and fellow citizens. They may take pride in their personal habits or in having the smallest garbage can on the block. Or they may have excellent services and think that is just how things are meant to be, rather than recognizing that they hit the jackpot of privilege in disparately distributed basic services that are tied to a history of state-built segregation.[4] The ordinary continues to be a hidden site of state operation and a construct of political identity.

Although there are good methodological reasons to select an issue like garbage collection and disposal—the number of cases and variation in outcomes—there are good political reasons too. Politics does not take place only, or even mostly, in the capital of Washington, DC, and people do not participate only, or even mostly, in politics episodically every four years around a presidential election. Instead, politics takes place in the locations where Americans live and work. They participate in it day in and day out (see figure 8.1). Politics is so ordinary and ingrained in American life that we do not see it anymore. It is hiding in plain sight.[5]

Politics is fundamentally about power. It's all too easy to look for power in contested issues and to see who wins (elections, legislation, court cases) rather than looking for power in issues that are not contested and may not be seen as political at all.[6] As the history of garbage collection and disposal shows, garbage was once extremely political. The issue was at the forefront of late-nineteenth-century political campaigns, and it led to the fall of mayoral administrations. What we think of and measure as political today—creating ordinances and designing programs—was the comparatively *easy* part. City officials also had to marshal resources, induce compliance, and deflect blame. The fact that Americans take their trash out to the designated place, at the designated time, in the designated receptacle and think of it merely as an individual habit (rather than an exercise of power) is a significant achievement of all of those governments.

We chose to examine the development of garbage collection and disposal programs, an issue that almost every American participates in but few recognize the politics of. Others, however, can apply the same analysis to any public problem or any ordinary program—lighting, sewers, schools, land use—and ask: Are there services?[7] Who delivers them? On what resources do they rely? And with what effect? Our strategy in answering these questions was to map the actors who play a key role and to analyze the resources they use to achieve their objective.

The purchase in recognizing the politics and power in everyday practice lies in tracking the mechanisms of governing. We can recognize how they build capacity and how they shape and interact with people's lives. Mapping everyday politics in practice means looking for the relations that government fosters and the informal resources it enlists and determining whether social inequality is a resource for meeting the imperatives of governing.[8] It considers the long-term implications when these informal resources are used to build formal public programs.

San Francisco's scavengers, who were so deft at resisting takeover by the city, maintaining their practice of making contracts with householders, joined the city halfway in 1932, when a garbage ordinance provided for select scavengers to obtain permits from the director of public health. The Scavengers Protective Association and Sunset Scavenger Company emerged with their monopoly intact, eventually falling under the umbrella of Recology. City supervisors tried to rescind this arrangement in 2011 with a ballot measure, but voters overwhelmingly voted it down, with 77 percent voting against change. Like the Scavengers Protective Association in the 1920s, the company emerged with its monopoly complete.[9]

In 2020 Recology was caught in a scandal when a group of executives was implicated in a scheme to bribe the director of the department of public works. The bribes included cash funneled through nonprofits affiliated with the director, but, as with the corruption of the nineteenth century, it also included very particular perks, such as landing a job for the director's son, or a two-night stay in a luxury New York City hotel for public works employees who were in town to tour a pneumatic waste collection system. And as with the corruption of old, Recology was able to recover, making a deal with the Department of Justice and mollifying the public with a $100 million reimbursement rate agreement.[10] Notably, "Recology's operations in San Francisco remain largely unchanged, given its lock on most of the city's collection."[11] The people, in fact, remain loyal to their collectors, whom they distinguish from management. In 1929 the City Sanitation Guild asked residents to stick with their garbage man. They have. The garbage collectors have maintained relationships with household residents, utilizing a resource of governance. The legacy of the Sunset Scavenger Company maintains its hold on garbage collection in San Francisco.

NOTES

Works frequently cited have been identified by the following abbreviations:

ASC   Archives Service Center, University of Pittsburgh
CCPL   The Charleston Archive, Charleston County Public Library
HPGTC Historic Pittsburgh General Text Collection, University of
       Pittsburgh, https://historicpittsburgh.org/
HSWP   Detre Library and Archives, Senator John Heinz
       History Center, Historical Society of Western Pennsylvania,
       Pittsburgh
HTDL   Hathi Trust Digital Library, https://www.hathitrust.org/
LFPL   Louisville Free Public Library
MSA   Missouri State Archives, Jefferson City
NOPL   City Archives & Special Collections, New Orleans Public
       Library
SCHS   South Carolina Historical Society
SFPL   San Francisco History Center, San Francisco Public Library
THNOC The Historic New Orleans Collection, Williams Research
       Center

Introduction

1. Lily Baum Pollans, "Greening Infrastructural Services: The Case of Waste Management Service in San Francisco," MIT Department of Urban Studies and Planning, December 2012, 7, http://web.mit.edu/nature/projects_12/pdfs/Pollans_SFwaste_2012.pdf

2. "Over 100 Years of Service," Recology, https://www.recology.com/about-us/our-history/.

3. Charles V. Chapin, *Municipal Sanitation in the United States* (Providence: Snow & Farnham, 1901), 697.

4. Chapin, *Municipal Sanitation,* 688–94.

5. Chapin, *Municipal Sanitation,* 730; New York Bureau of Municipal Research, *The City of Pittsburgh, Pennsylvania: On a Survey of the Department of Public Health, et al.* (Pittsburgh: Pittsburgh Printing, 1913), 30, HPGTC.

6. A privy was a buried hole or container to hold human waste. Paul Underwood Kellogg, ed., *The Pittsburgh District: Civic Frontage, The Pittsburgh Survey* (New York: Survey Associates, 1914), 100–101, Russell Sage Foundation,

https://www.russellsage.org/sites/all/files/Kellogg_The%20Pittsburgh%20 District_0.pdf.

7. Kellogg, *The Pittsburgh District*, 136. See Chapin, *Municipal Sanitation*, 172.

8. John Duffy, *The Sanitarians: A History of American Public Health* (Urbana: University of Illinois Press, 1990), 77, 175–82.

9. Chadwick Montrie, "A Path to Reform: Confronting the Garbage Crisis in Louisville, 1865–1873," *Filson Club History Quarterly* 70, no. 1 (January 1996): 29; "Sanitarians' Council Talk Over the South Side Typhoid Fever Epidemic," *Pittsburgh Daily Post,* May 1, 1880, 4; Reverend Hugh Miller Thompson, "Method Introduced by the Auxiliary Sanitary Association for Disposing of the Garbage of New Orleans," *Public Health Papers and Reports* 5 (1879): 33.

10. See Martin V. Melosi, *Garbage in the Cities: Refuse, Reform, and the Environment* (Pittsburgh: University of Pittsburgh Press, 2005); William McGowan, "America's Wasteland: A History of America's Garbage Industry, 1880–1989," *Business and Economic History* 24, no. 1 (Fall 1995): 155–63; Stanley K. Schultz and Clay McShane, "To Engineer the Metropolis: Sewers, Sanitation, and City Planning in Late-Nineteenth-Century America," *Journal of American History* 65, no. 2 (September 1978): 389–411; Joel A. Tarr, *The Search for the Ultimate Sink: Urban Pollution in Historical Perspective* (Akron: University of Akron Press, 1996); Carl Zimring, *Cash for Your Trash: Scrap Recycling in America* (New Brunswick: Rutgers University Press, 2009).

11. Melosi, *Garbage in the Cities,* 17.

12. Rudolph Hering and Samuel A. Greeley, *Collection and Disposal of Municipal Refuse* (New York: McGraw-Hill, 1921), 2.

13. F. C. Bamman, "War's Influence on the Garbage Pail," *Engineering News* 82, no. 8 (February 20, 1919): 373.

14. Bamman, "War's Influence," 374.

15. Chapin, *Municipal Sanitation;* Hering and Greeley, *Collection and Disposal of Municipal Refuse.*

16. Duffy, *The Sanitarians.*

17. David Cutler and Grant Miller, "The Role of Public Health Improvements in Health Advances: The Twentieth-Century United States," *Demography* 42, no. 1 (2005): 1–22.

18. John N. Collins and Bryan T. Downes, "The Effects of Size on the Provision of Public Services: The Case of Solid Waste Collection in Smaller Cities," *Urban Affairs Review* 12, no. 3 (1977): 333.

19. Amy Bridges, introduction to *Urban Citizenship and American Democracy,* ed. Amy Bridges and Michael Javen Fortner (Albany: SUNY Press, 2016), 1–21.

20. Kimberley S. Johnson, "The Color Line and the State: Race and American Political Development," in *The Oxford Handbook of American Political Development,* ed. Richard Valelly, Suzanne Mettler, and Robert C. Lieberman (Oxford: Oxford University Press, 2016), 593–624. See also Eric H. Monkkonen, *America Becomes Urban: The Development of U.S. Cities and Towns, 1780–1980* (Berkeley:

University of California Press, 1988); William Novak, *The People's Welfare: Law and Regulation in Nineteenth-Century America* (Chapel Hill: University of North Carolina Press, 1996).

21. Carol Nackenoff and Julie Novkov, "Statebuilding in the Progressive Era: A Continuing Dilemma in American Political Development," in *Statebuilding from the Margins: Between Reconstruction and the New Deal*, ed. Carol Nackenoff and Julie Novkov (Philadelphia: University of Pennsylvania Press, 2014), 1.

22. Cutler and Miller, "The Role of Public Health Improvements in Health Advances." See also Daniel Sledge, "War, Tropical Disease, and the Emergence of National Public Health Capacity in the United States," *Studies in American Political Development* 26, no. 2 (2012): 125–62; Daniel Sledge, *Health Divided: Public Health and Individual Medicine in the Making of the Modern American State* (Lawrence: University Press of Kansas, 2017).

23. Jessica Trounstine, *Segregation by Design: Local Politics and Inequality in American Cities* (New York: Cambridge University Press, 2018).

24. Schultz and McShane, "To Engineer the Metropolis," 391. See also Monkkonen, *America Becomes Urban.*

25. Paul Manna, *School's In: Federalism and the National Education Agenda* (Washington, DC: Georgetown University Press, 2006), 31–33; Stephen Skowronek, *Building a New American State: The Expansion of National Administrative Capacities, 1877–1920* (Cambridge: Cambridge University Press, 1982).

26. Patricia Strach, Kathleen Sullivan, and Elizabeth Pérez-Chiqués, "The Garbage Problem: Corruption, Innovation, and Capacity in Four American Cities, 1890–1940," *Studies in American Political Development* 33, no. 2 (2019): 6.

27. "Report of the Committee on Disposal of Waste and Garbage," *Public Health Papers and Reports* 17 (1891): 90–143.

28. Col. W. F. Morse, "The Collection and Disposal of the Refuse of Large Cities," *Public Health Papers and Reports* 20 (1894): 187.

29. Rudolph Hering, "Report of the Committee on Disposal of Garbage and Refuse," *Public Health Papers and Reports* 29 (1903): 129–33.

30. American Public Health Association, "Report of the Committee on the Disposal of Garbage and Refuse," *Public Health Papers and Reports* 23 (1897): 208.

31. George E. Waring, *Street-Cleaning and the Disposal of a City's Waste* (New York: Doubleday & McClure, 1898).

32. Craig M. Brown and Charles N. Halaby, "Machine Politics in America, 1870–1945," *Journal of Interdisciplinary History* 17, no. 3 (1987): 587–612.

33. Valdimer Orlando Key Jr., "The Techniques of Political Graft in the United States: A Part of a Dissertation Submitted to the Faculty of the Division of the Social Sciences in Candidacy for the Degree of Doctor of Philosophy" (PhD diss., University of Chicago, 1936), 5.

34. The actors enlisted to solve the garbage can problem were overwhelmingly women, quite often white women in civic organizations. We use "gender" as the category, however, to target the construction of political identity in the

process of contributing. Those women who stepped forward relied on their class and race privilege to emerge as ideal homemakers who could serve as a model for others, thus relying on exclusionary mechanisms to construct their own political identity which became the standard. See Joan W. Scott, "Gender: A Useful Category of Historical Analysis," *American Historical Review* 91, no. 5 (December 1986): 1053–75; Rose McDermott and Peter K. Hatemi, "Distinguishing Sex and Gender," *PS: Political Science and Politics* 44, no. 1 (January 2011): 89–92.

35. Susan Strasser, *Waste and Want: A Social History of Trash* (New York: Holt, 1999); Suellen Hoy, *Chasing Dirt: The American Pursuit of Cleanliness* (New York: Oxford University Press, 1995); Angela Gugliotta, "Class, Gender, and Coal Smoke: Gender Ideology and Environmental Injustice in Pittsburgh, 1868–1914," *Environmental History* 5, no. 2 (April 2000): 165–93.

36. Michael Mann, "The Autonomous Power of the State: Its Origins, Mechanisms and Results," *European Journal of Sociology* 25, no. 2 (1984): 185–213.

37. Gabriel Rosenberg, *The 4-H Harvest: Sexuality and the State in Rural America* (Philadelphia: University of Pennsylvania Press, 2015).

38. See David Sibley, *Geographies of Exclusion: Society and Difference in the West* (London: Routledge, 1995).

39. See, for example, Natalia Molina, *Fit to Be Citizens? Public Health and Race in Los Angeles, 1879–1939* (Berkeley: University of California Press, 2006).

40. Carl Zimring, "Dirty Work: How Hygiene and Xenophobia Marginalized the American Waste Trades, 1870–1930," *Environmental History* 9, no. 1 (2004): 80.

41. Carl A. Zimring, *Clean and White: A History of Environmental Racism in the United States* (New York: New York University Press, 2015), 117.

42. Laurence Glasco, "Double Burden: The Black Experience in Pittsburgh," in *City at the Point: Essays on the Social History of Pittsburgh,* ed. Samuel Hays (Pittsburgh: University of Pittsburgh Press, 1987), 73–74.

43. Susan L. Smith, *Sick and Tired of Being Sick and Tired: Black Women's Health Activism in American, 1890–1950* (Philadelphia: University of Pennsylvania Press, 1995), 31–32.

44. W. E. Burghardt Dubois, ed., *The Negro American Family* (Atlanta: Atlanta University Press, 1908), 58–62.

45. Robin Nagle, *Picking Up: On the Streets and Behind the Trucks with the Sanitation Workers of New York City* (New York: Farrar, Straus, and Giroux, 2014).

46. Steve Estes, "'I am a Man!': Race, Masculinity, and the 1968 Memphis Sanitation Strike," *Labor History* 41, no. 2 (2000): 153–70.

47. Julie Novkov, "Rethinking Race in American Politics," *Political Research Quarterly* 61, no. 4 (December 2008): 650; Edmund Fong, "Reconstructing the 'Problem' of Race," *Political Research Quarterly* 61, no. 4 (December 2008): 660–70.

48. Laura E. Gómez, "Looking for Race in All the Wrong Places," *Law & Society Review* 46, no. 2 (June 2012): 234; Naomi Murakawa and Katherine

Beckett, "The Penology of Racial Innocence: The Erasure of Racism in the Study and Practice of Punishment," *Law & Society Review* 44, no. 3–4 (September–December 2010): 707.

49. Laura E. Gómez, "Understanding Law and Race as Mutually Constitutive: An Invitation to Explore an Emerging Field," *Annual Review of Law and Social Science* 6 (2010): 487–505.

50. Edward Morris, "Researching Race: Identifying a Social Construction through Qualitative Methods," *Symbolic Interaction* 30, no. 3 (Summer 2007): 409–25.

51. Johnson, "The Color Line and the State," 597.

52. See Suzanne Mettler, *Dividing Citizens: Gender and Federalism in New Deal Public Policy* (Ithaca: Cornell University Press, 1998); Robert C. Lieberman, *Shifting the Color Line: Race and the American Welfare State* (Cambridge: Harvard University Press, 1998).

53. Chapin, *Municipal Sanitation*, 670.

1. A Conceptual Roadmap

1. Patricia Strach and Kathleen S. Sullivan, "The State's Relations: What the Institution of Family Tells Us about Governance," *Political Research Quarterly* 64, no. 1 (2011): 94–106.

2. For example, see Kenneth N. Bickers and Robert M. Stein, "The Electoral Dynamics of the Federal Pork Barrel," *American Journal of Political Science* 40, no. 4 (1996): 1300–1326; David Mayhew, *Party Loyalty among Congressmen* (Cambridge: Harvard University Press, 1966); Daniel Carpenter, *Reputation and Power: Organizational Image and Pharmaceutical Regulation at the FDA* (Princeton: Princeton University Press, 2010).

3. For example, see Carol Nackenoff and Julie Novkov, "Statebuilding in the Progressive Era: A Continuing Dilemma in American Political Development," in *Statebuilding from the Margins: Between Reconstruction and the New Deal*, ed. Carol Nackenoff and Julie Novkov (Philadelphia: University of Pennsylvania Press, 2014), 1–31; Jessica Wang, "Dogs and the Making of the American State: Voluntary Association, State Power, and the Politics of Animal Control in New York City, 1850–1920," *Journal of American History* 98, no. 4 (2012): 998–1024; Kimberley S. Johnson, *Governing the American State: Congress and the New Federalism, 1877–1929* (Princeton: Princeton University Press, 2007).

4. See Lester M. Salamon, "The New Governance and the Tools of Public Action: An Introduction," in *Tools of Government: A Guide to the New Governance*, ed. Lester M. Salamon (New York: Oxford University Press, 2002), 8.

5. For a summary, see Christopher Ansell and Jacob Torfing, *Handbook on Theories of Governance* (Northampton, MA: Edward Elgar Publishing, 2016); Jon Pierre and B. Guy Peters, *Governance, Politics, and the State* (London: Red Globe Press, 2020).

6. Carol Nackenoff and Julie Novkov, eds., *Statebuilding from the Margins: Between Reconstruction and the New Deal* (Philadelphia: University of Pennsylvania Press, 2014).

7. Pierre and Peters, *Governance*, 12.

8. Nackenoff and Novkov, *Statebuilding from the Margins;* James Sparrow, William Novak, and Stephen Sawyer, eds., *Boundaries of the State in US History* (Chicago: University of Chicago Press, 2015).

9. Desmond S. King and Rogers M. Smith, "'Without Regard to Race': Critical Ideational Development in Modern American Politics," *Journal of Politics* 76, no. 4 (July 2014): 958–71; Rogers Smith, *Civic Ideals: Conflicting Visions of Citizenship in U.S. History* (New Haven: Yale University Press, 1999).

10. Christopher Howard, *The Hidden Welfare State: Tax Expenditures and Social Policy in the United States* (Princeton: Princeton University Press, 1997); Marie Gottschalk, *The Shadow Welfare State: Labor, Business, and the Politics of Health Care in the United States* (Ithaca: Cornell University Press, 2000); Suzanne Mettler, *The Submerged State: How Invisible Government Policies Undermine American Democracy* (Chicago: University of Chicago Press, 2011); Patricia Strach, *All in the Family: The Private Roots of American Public Policy* (Stanford: Stanford University Press, 2007); Suzanne Mettler, *Dividing Citizens: Gender and Federalism in New Deal Public Policy* (Ithaca: Cornell University Press, 1998); Kimberley S. Johnson, *Governing the American State: Congress and the New Federalism, 1877–1929* (Princeton: Princeton University Press, 2007); Desmond King and Robert C. Lieberman, "Finding the American State: Transcending the 'Statelessness' Account," *Polity* 40, no. 3 (2008): 368–78.

11. Chloe Thurston, "Black Lives Matter, American Political Development, and the Politics of Visibility," *Politics, Groups and Identities* 6, no. 1 (2018): 167.

12. Joseph S. Nye, "Corruption and Political Development: A Cost-Benefit Analysis," *American Political Science Review* 61, no. 2 (June 1967): 419.

13. Valdimer Orlando Key Jr., "The Techniques of Political Graft in the United States: A Part of a Dissertation Submitted to the Faculty of the Division of the Social Sciences in Candidacy for the Degree of Doctor of Philosophy" (PhD diss., University of Chicago, 1936), 5.

14. James C. Scott writes that "whereas party manifestos, general legislation, and policy declarations are the formal façade of the political structure, corruption stands in sharp contrast to these features as an informal political system in its own right." James C. Scott, "Handling Historical Comparisons Cross-Nationally," in *Political Corruption: A Handbook,* ed. Arnold Heidenheimer, Michael Johnston, and Victor T. LeVine (New Brunswick, NJ: Transaction Publishers, 1999), 129.

15. For examples of political power of machine governments, see Samuel P. Hays, *Conservation and the Gospel of Efficiency: The Progressive Conservation Movement, 1890–1920* (Pittsburgh: University of Pittsburgh Press, 1999); Richard Hofstadter, *The Age of Reform: From Bryan to FDR* (New York: Knopf, 1955); James C. Scott, "Corruption, Machine Politics, and Political Change," *American Political Science Review* 63, no. 4 (1969): 1142–58. For limitations of machines and the literature about them, see Clarence Stone, *Regime Politics: Governing Atlanta, 1946–1988* (Lawrence: University Press of Kansas, 1989); Jon C. Teaford, "Finis for Tweed and Steffens: Rewriting the History of Urban Rule," *Reviews in American History* 10 (1982): 133–49; Kenneth Finegold, *Experts and Politicians: Reform Challenges to Machine Politics in New York, Cleveland, and Chicago*

(Princeton: Princeton University Press, 1995); Jessica Trounstine, *Political Monopolies in American Cities: The Rise and Fall of Bosses and Reformers* (Chicago: University of Chicago Press, 2008); Amy Bridges, *A City in the Republic: Antebellum New York and the Origins of Machine Politics* (New York: Cambridge University Press, 1984); Richard Schneirov, "Urban Regimes and the Policing of Strikes in Two Gilded Age Cities: New York and Chicago," *Studies in American Political Development* 33, no. 2 (2019): 258–74.

16. See, for example, Christopher K. Ansell and Arthur L. Burris, "Bosses of the City Unite! Labor Politics and the Political Machine Consolidation, 1870–1910," *Studies in American Political Development* 11, no. 1 (1997): 1–43; Steven Erie, *Rainbow's End: Irish-Americans and the Dilemmas of Urban Machine Politics, 1840–1985* (Berkeley: University of California Press, 1988).

17. Amy Bridges, *Morning Glories: Municipal Reform in the Southwest* (Princeton: Princeton University Press, 1997).

18. Bo Rothstein and Aiysha Varraich, *Making Sense of Corruption* (New York: Cambridge University Press, 2017).

19. Patricia Strach, Kathleen Sullivan, and Elizabeth Pérez-Chiqués. "The Garbage Problem: Corruption, Innovation, and Capacity in Four American Cities, 1890–1940," *Studies in American Political Development* 33, no. 2 (2019): 209–33.

20. Colin Gordon, *Citizen Brown: Race, Democracy, and Inequality in the St. Louis Suburbs* (Chicago: University of Chicago Press, 2019); Joseph Lowndes et al., eds., *Race and American Political Development* (New York: Routledge, 2008); Robert Lieberman, *Shifting the Color Line: Race and the American Welfare State* (Cambridge: Harvard University Press, 2001); Mettler, *Dividing Citizens*; Jessica Trounstine, *Segregation by Design: Local Politics and Inequality in American Cities* (New York: Cambridge University Press, 2018).

21. Maureen A. Flanagan, "Gender and Urban Political Reform: The City Club and the Woman's City Club of Chicago in the Progressive Era," *American Historical Review* 95, no. 4 (1990): 1048; Elizabeth Belanger, "The Neighborhood Ideal: Local Planning Practices in Progressive-Era Women's Clubs," *Journal of Planning History* 8, no. 2 (May 2009): 87–110; Angela Gugliotta, "Class, Gender, and Coal Smoke: Gender Ideology and Environmental Injustice in Pittsburgh, 1868–1914," *Environmental History* 5, no. 2 (April 2000):174.

22. W. E. Burghardt Dubois, ed., *The Negro American Family* (Atlanta: Atlanta University Press, 1909), 58–62; Laurence Glasco, "Double Burden: The Black Experience in Pittsburgh," in *City at the Point: Essays on the Social History of Pittsburgh*, ed. Samuel Hays (Pittsburgh: University of Pittsburgh Press, 1987), 73–74; Susan L. Smith, *Sick and Tired of Being Sick and Tired: Black Women's Health Activism in America, 1890–1950* (Philadelphia: University of Pennsylvania Press, 1995), 31–32; Carl Zimring, "Dirty Work: How Hygiene and Xenophobia Marginalized the American Waste Trades, 1870–1930," *Environmental History* 9, no. 1 (2004): 80; Carl A. Zimring, *Clean and White: A History of Environmental Racism in the United States* (New York: New York University Press, 2015), 117.

23. Richard F. Bensel, *Yankee Leviathan: The Origins of Central State Authority in America, 1859–1877* (Cambridge: Cambridge University Press, 1990); Daniel P.

Carpenter, *The Forging of Bureaucratic Autonomy: Reputations, Networks, and Policy Innovation in Executive Agencies, 1862–1928* (Princeton: Princeton University Press, 2002); Karen Orren and Stephen Skowronek, *The Search for American Political Development* (Cambridge: Cambridge University Press, 2004); Stephen Skowronek, *Building a New American State: The Expansion of National Administrative Capacities, 1877–1920* (Cambridge: Cambridge University Press, 1982).

24. Carpenter, *The Forging of Bureaucratic Autonomy.*

25. Orren and Skowronek, *The Search for American Political Development,* 20.

26. Richardson Dilworth, introduction to *The City in American Political Development,* ed. Richardson Dilworth (New York: Routledge, 2009), 1–13; Desmond S. King and Rogers M. Smith, "Racial Orders in American Political Development," *American Political Science Review* 99, no. 1 (February 2005): 75–92; Robert C. Lieberman, "Ideas, Institutions, and Political Order: Explaining Political Change," *American Political Science Review* 96, no. 4 (December 2002): 697–712.

27. Richardson Dilworth and Timothy Weaver, *How Ideas Shape Urban Political Development* (Philadelphia: University of Pennsylvania Press, 2020).

28. John Kingdon, *Agendas, Alternatives, and Public Policies* (New York: Longman Press, 2003), 1.

29. Trounstine, *Political Monopolies in American Cities.*

30. City of Pittsburgh, Annual Report of the Departments and Offices, Report of the City Controller, 1915, 91, HSWP.

31. For more on the importance of comparative method and primary fieldwork in waste studies, see Raul Pacheco-Vega, "Comparative Methods for the Study of Waste," in *The Routledge Handbook of Waste Studies,* ed. Zsuzsa Gille and Josh Lepawsky (London: Routledge, 2021), 125.

32. Lily Baum Pollans, *Resisting Garbage: The Politics of Waste Management in American Cities* (Austin: University of Texas Press, 2021), 9.

33. Regional differences in municipal programs likely reflect regional differences in structure, which themselves can be traced back to class or group control of the rules of governing. See Amy Bridges and Richard Kronick, "Writing the Rules to Win the Game: The Middle-Class Regimes of Municipal Reformers," *Urban Affairs Review* 34 (May 1999): 691–706.

34. While our focus is on the early decades of garbage collection, we occasionally move beyond this period, notably in Pittsburgh, Louisville, and Birmingham, to track the end points of patterns we see in this earlier period.

2. Ready to Help

1. US Bureau of the Census, table 10, "Population of the 100 Largest Urban Places: 1870," and table 12, "Population of the 100 Largest Urban Places: 1890," https://www.census.gov/library/working-papers/1998/demo/POP-twps0027.html.

2. George A. Soper, *Modern Methods of Street Cleaning* (New York: The Engineering News Publishing Company, 1909), 8.

3. "The South Side Scourge: Investigation by the Board of Health Committee," *Pittsburgh Daily Post,* April 5, 1880, 4.

4. "Sanitarians' Council Talk Over the South Side Typhoid Fever Epidemic," *Pittsburgh Daily Post*, May 1, 1880, 4.

5. "Health Officer's Report," *San Francisco Municipal Reports for the Fiscal Year 1871–72* (San Francisco: Cosmopolitan Printing Co., 1872), 216.

6. "Annual Report of the Harbor and Wharf Commissioner," *The Mayor's Message with Accompanying Documents to the Municipal Assembly of the City of Saint Louis* (St. Louis: Wm. Bieginger & Co., 1885), 294.

7. For an account of waste more generally, see Joel A. Tarr, *The Search for the Ultimate Sink: Urban Pollution in Historical Perspective* (Akron: University of Akron Press, 1996); Martin V. Melosi, *Garbage in the Cities: Refuse, Reform, and the Environment*, rev. ed. (Pittsburgh: University of Pittsburgh Press, 2005).

8. Mira Engler, "Repulsive Matter: Landscapes of Waste in the American Middle-Class Residential Domain," *Landscape Journal* 16, no. 1 (Spring 1997): 60–79; Angela Gugliotta, "How, When, and for Whom Was Smoke a Problem for Pittsburgh?" in *Devastation and Renewal: An Environmental History of Pittsburgh and Its Region*, ed. Joel A. Tarr (Pittsburgh: University of Pittsburgh Press, 2003), 110–25; Adam Rome, "Coming to Terms with Pollution: The Language of Environmental Reform, 1865–1917," *Environmental History* 1, no. 3 (July 1996): 6–28; Deborah Stone, *Policy Paradox: The Art of Political Decision Making* (New York: Norton, 2012).

9. Tarr, *The Search for the Ultimate Sink*, 10–11.

10. John Duffy, *The Sanitarians: A History of American Public Health* (Urbana: University of Illinois Press, 1992), 94.

11. Charles E. Rosenberg, *The Cholera Years: The United States in 1832, 1849, and 1866* (Chicago: University of Chicago Press, 1987); Margaret Humphreys, *Yellow Fever and the South* (Baltimore: Johns Hopkins University Press, 1999).

12. Jon A. Peterson, "The Impact of Sanitary Reform upon American Urban Planning," *Journal of Social History* 13, no. 1 (1979): 83–103.

13. APHA Constitution, cited in Duffy, *The Sanitarians*, 130.

14. Jerrold Michael, "The National Board of Health, 1879–1883," *Public Health Reports* 126, no. 1 (January–February 2011): 127; See also Dennis East II, "Health and Wealth: Goals of the New Orleans Public Health Movement, 1879–84," *Louisiana History* 9, no. 3 (1968): 249.

15. Rudolph Hering and Samuel A. Greeley, *Collection and Disposal of Municipal Refuse* (New York: McGraw-Hill, 1921), 2.

16. Rudolph Hering, "Modern Practice in the Disposal of Refuse," *Journal of the American Public Health Association* 1, no. 12 (December 1911): 910.

17. Hering and Greeley, *Collection and Disposal*, 29.

18. Hering, "Modern Practice," 910.

19. John Ellis, *Yellow Fever and Public Health in the New South* (Lexington: University Press of Kentucky, 1992), 34.

20. Duffy, *The Sanitarians*, 37–39, 41–42, 60.

21. Jo Ann Carrigan, "Yellow Fever in New Orleans, 1853: Abstractions and Realities," *Journal of Southern History* 25, no. 3 (1959): 342, 345.

22. William Blackstone, *Commentaries on the Laws of England*, bk. 3 (Oxford: Clarendon Press, 1765–1769), 217, 216.

23. Charles V. Chapin, *Municipal Sanitation in the United States* (Providence: Snow & Farnham, 1901), 143–65.

24. John Fabian Witt, *The Accidental Republic: Crippled Workingmen, Destitute Widows, and the Remaking of American Law* (Cambridge: Harvard University Press, 2004).

25. See Karen Orren, *Belated Feudalism: Labor, the Law, and Political Development in the United States* (New York: Cambridge University Press, 1999); Stephen Skowronek, *Building a New American State: The Expansion of National Administrative Capacities, 1877–1920* (New York: Cambridge University Press, 1982); Noga Morag-Levine, *Chasing the Wind: Regulating Air Pollution in the Common Law State* (Princeton: Princeton University Press, 2003).

26. Humphreys, *Yellow Fever*, 13.

27. Quoting Dr. J. F. Kennedy, "The Waring System of Sewerage," *Paving and Municipal Engineering* 1, no. 1 (June 1890): 4. Kennedy was secretary of the Iowa State Board of Health.

28. *A Digest of the Acts of Assembly relating to and the general ordinances of the city of Allegheny from 1840 to April 1st, 1897* (Pittsburgh: Pittsburgh Printing Co., 1897), 140, HPGTC.

29. *Report of the Sanitary Commission of New Orleans on the Epidemic Yellow Fever* (New Orleans: Picayune Office, 1854), 390, 472.

30. Margaret Humphreys, *Yellow Fever and the South* (Baltimore: Johns Hopkins University Press, 1992), 54.

31. East, "Health and Wealth," 249.

32. Melosi, *Garbage in the Cities*, 18.

33. Craig E. Colten, *An Unnatural Metropolis: Wresting New Orleans from Nature* (Baton Rouge: Louisiana State University Press, 2005), 58.

34. Hugh Miller Thompson, "Method Introduced by the Auxiliary Sanitary Association for Disposing of the Garbage in New Orleans," *Public Health Papers and Reports* 5 (1879): 33–34; Edward Fenner, *Annual Address at the Regular Meeting of the New Orleans Auxiliary Sanitary Association, November 23, 1880* (New Orleans: New Orleans Democrat Office, 1880).

35. Joy Jackson, *New Orleans in the Gilded Age: Politics and Urban Progress, 1880–1896* (New Orleans: Louisiana Historical Association, 1987), 159–60.

36. "Annual Report of the Health Commissioner," *Eighth Annual Report of the Health Commissioner, City of St. Louis, for the Fiscal Year Ending April 13, 1885*, 323, MSA.

37. "Annual Report of the Health Commissioner," *Eighth Annual Report*, 9, MSA.

38. Chapin, *Municipal Sanitation*. C.-E. A. Winslow and P. Hansen followed up Chapin's study by sending surveys to 161 cities with a population over 25,000, to focus on the particulars of garbage collection and disposal practices so as to determine if garbage should be sorted and how it should be treated. "Some Statistics of Garbage Disposal for the Larger American Cities in 1902," *Public Health Papers and Reports* 29 (Columbus: Fred J. Heer, 1904): 141–65.

39. George E. Waring, *Street-Cleaning and the Disposal of a City's Wastes* (New York: Doubleday & McClure, 1898).

40. Alexis de Tocqueville, *Democracy in America*, ed. and trans. Harvey C. Mansfield and Delba Winthrop (Chicago: University of Chicago Press, 2000), 489–92.

41. Robert Putnam, *Bowling Alone: The Collapse and Revival of American Community* (New York: Simon and Schuster, 2000).

42. Theda Skocpol and Jennifer Lynn Oser, "Organization Despite Adversity: The Origins and Development of African American Fraternal Associations," *Social Science History* 28, no. 3 (Fall 2004): 367–437; Gerda Lerner, "Early Community Work of Black Club Women," *Journal of Negro History* 59, no. 2 (April 1974): 158–67.

43. Thompson, "Method Introduced by the Auxiliary Sanitary Association," 33; Ellis, *Yellow Fever*, 93–94.

44. Ellis, *Yellow Fever*, 95.

45. Robert Williams, "Martin Behrman and New Orleans Civic Development, 1904–1920," *Louisiana History* 2, no. 4 (Autumn 1961): 394.

46. Beverly Jones, "Mary Church Terrell and the National Association of Colored Women, 1896 to 1901," *Journal of Negro History* 67, no. 1 (Spring 1982): 22.

47. *A History of the Club Movement among the Colored Women of the United States of America, as Contained in the Minutes of the Convention, Held in Boston, July 29, 30, 31, 1895, and of the National Federation of Afro-American Women, Held in Washington, D.C., July 20, 21, 22, 1896* (Washington, DC: National Association of Colored Women's Clubs, 1902), 104–5; Records of the National Association of Colored Women's Clubs, 1895–1992, pt. 1, Minutes of National Conventions, Publications, and President's Office Correspondence, https://congressional.proquest.com/histvault?q=001554-001-0001&accountid=12954.

48. *History of the Club Movement*, 32, 78.

49. *History of the Club Movement*, 47, 51.

50. *History of the Club Movement*, 104.

51. "Afro-American Notes," *Pittsburgh Press*, December 6, 1908, 39.

52. Angela Gugliotta, "Class, Gender, and Coal Smoke: Gender Ideology and Environmental Injustice in Pittsburgh, 1868–1914," *Environmental History* 5 (April 2000): 165–93.

53. Gugliotta, "How, When, and for Whom Was Smoke a Problem for Pittsburgh?" 115–16.

54. "Will Pave Alleys: Mayor Kennedy Will Insist on the Lanes Being Improved," *Pittsburgh Commercial Gazette*, April 24, 1895, 8.

55. "Pretty Hard Job Tackled: Civic Club Organized to Fight Prevalent Municipal Corruption Hard on the Local System," *Pittsburgh Daily Post*, October 8, 1895, 1.

56. "The Index: Celebrated Twentieth Anniversary Has Accomplished Much," November 13, 1915, Clippings—Civic Club History, 1902–1915, Civic Club of Allegheny County Records, 1896–1974, AIS.1970.02, ASC.

57. Civic Club of Allegheny County, *Fifteen Years of Civic History* (Pittsburgh: Nicholson Printing Co., 1910), 13–14, HPGTC.

58. M. N. Baker, ed., *Municipal Year Book* (New York: Engineering News Publishing Co., 1902), xxvii, xiii.

59. Chapin, *Municipal Sanitation*, 172, 252, 264.

60. Chapin, *Municipal Sanitation*, 262; Hering and Greeley, *Collection and Disposal*, 3.

61. Edwin Fisher, "President's Annual Address," *Municipal Engineering* 23, no. 5 (November 1902): 332.

62. Baker, *Municipal Yearbook*, xiii.

63. Hering and Greeley, *Collection and Disposal*, 105.

64. "The Prospects for Young Engineers: A Symposium," *Engineering News* 32, no. 1 (July 5, 1894): 5.

65. Stanley K. Schultz and Clay McShane, "To Engineer the Metropolis: Sewers, Sanitation, and City Planning in Late-Nineteenth-Century America," *Journal of American History* 65, no. 2 (September 1978): 401.

66. Thos. D. DeVilbis, "Collection and Disposal of Garbage," *Proceedings of the Second Annual Convention of the American Society of Municipal Improvements* (Cincinnati: Commercial Gazette, 1895), 6, 83–100.

67. Schultz and McShane, "To Engineer the Metropolis."

68. "The Future of the American Society of Municipal Improvements," *Engineering News* 44, no. 10 (September 6, 1900): 164–65.

69. "Report of the Committee on Disposition of Garbage and Street Cleaning," *Engineering News* 44, no. 10 (September 6, 1900): 172.

70. "The Latest Garbage Disposal Statistics," *Engineering News* 50, no. 19 (November 5, 1903): 413.

71. "Needed Reforms in the Collection and Disposal of City Refuse," *Engineering News* 59, no. 17 (April 23, 1908): 462.

72. "Beginning," *Paving and Municipal Engineering* 1, no. 1 (June 1890), 3.

73. "Outline of the Proceedings," *Municipal Engineering* 11, no. 5 (November 1896): 279.

74. "Garbage Disposal and Apparatus Therefor," *Municipal Engineering* 21, no. 3 (September 1901): 152; "Garbage Disposal, Street Cleaning, and Sprinkling," *Municipal Engineering* 21, no. 5 (November 1901): 376.

75. "Recent Inventions," *Municipal Engineering* 23, no. 1 (July 1902): 61; "Recent Inventions," *Municipal Engineering* 23, no. 3 (September 1902): 219.

76. "Garbage and Refuse Wagons," *Municipal Engineering* 21, no. 5 (November 1901): 332.

77. "Collection and Incineration of Garbage in Sewickley, Pa.," *Municipal Engineering* 52, no. 2 (February 1917): 59.

78. "The Garbaget," *Municipal Engineering* 52, no. 1 (January 1917): 34.

79. "Disposal of Garbage and Refuse in Seattle," *Municipal Engineering* 48, no. 2 (February 1915): 118.

80. "Modern Methods of Refuse and Garbage Disposal," *Municipal Engineering* 52, no. 6 (June 1917): 318.

81. Edwin Fisher, "President's Annual Address," *Municipal Engineering* 23, no. 5 (November 1902): 333.

82. Hering and Greeley, *Collection and Disposal of Municipal Refuse*, 3.

83. M. N. Baker, *Municipal Engineering and Sanitation* (New York: Macmillan Company, 1906), 3–4, 5–6, 164–65, 157, 160.

84. Baker, *Municipal Engineering and Sanitation*, 165, 238–39.

85. "Public Health and Municipal Government," *Annals of the American Academy of Political and Social Science* 1, no. 1, suppl. (1891): 6, 7.

86. *Proceedings of the National Conference for Good City Government held at Philadelphia, January 25 and 26, 1894* (Philadelphia: Municipal League, 1894), 46.

87. Clinton Rogers Woodruff, "The National Municipal League," *Proceedings of the American Political Science Association* 5 (1908): 132.

88. Clinton Rogers Woodruff, "A Year's Disclosure and Development," *Pamphlet No. 11. Publications of the National Municipal League* (Philadelphia: National Municipal League, 1904).

89. John S. Billings, "Good City Government from the Standpoint of the Physician and Sanitarian," in *Proceedings of the Second National Conference for Good City Government*, vols. 2–3 (Philadelphia: National Municipal League, 1895), 492–99.

90. George A. Soper, "The Work of Boards of Health," *Popular Science Monthly* 74 (March 1909): 238.

91. *Proceedings of the National Conference of Good City Government and the Fourteenth Annual Meeting of the National Municipal League* (Philadelphia: National Municipal League, 1908), 11.

3. Ready to Profit

1. Valdimer Orlando Key Jr., "The Techniques of Political Graft in the United States: A Part of a Dissertation Submitted to the Faculty of the Division of the Social Sciences in Candidacy for the Degree of Doctor of Philosophy" (PhD diss., University of Chicago, 1936), 5; Patricia Strach, Kathleen Sullivan, and Elizabeth Pérez-Chiqués. "The Garbage Problem: Corruption, Innovation, and Capacity in Four American Cities, 1890–1940." *Studies in American Political Development* 33, no. 2 (2019): 209–33.

2. Lana Stein, *St Louis Politics: The Triumph of Tradition* (St. Louis: Missouri Historical Society Press, 2002), 3.

3. Truman Port Young, "The Scheme of Separation of City and County Governments in Saint Louis—Its History and Purposes," *Proceedings of the American Political Science Association, Eighth Annual Meeting* (1911): 97–108.

4. Quoted in John D. Lawson, *American State Trials* vol. 9 (St. Louis: F. F. Thomas Law Books Co., 1918), 493.

5. Alexander Scot McConachie, "The 'Big Cinch': A Business Elite in the Life of a City, Saint Louis, 1895–1915" (PhD diss., Washington University, 1977), 169.

6. "A Delegate's Divvy," *St. Louis Post-Dispatch*, April 4, 1883, 3.

7. "Col. Ed. Butler's Remarkable Career and His Own Terse Story of His Life," *St. Louis Post-Dispatch*, September 10, 1911, 18, cited in Lawson, *American State Trials*, 9:493–94; "A Delegate's Divvy."

8. Cited but with no credit given in Works Progress Administration, Project 665-64-3-112, "Biographies of the Mayors of New Orleans," May 1939, 130, NOPL.

9. Edward F. Haas, "John Fitzpatrick and Political Continuity in New Orleans, 1896–1899," *Louisiana History* 22, no. 1 (1981): 10.

10. City of New Orleans, Pay Roll, Department of Public Works, 1883, NOPL.

11. John Duffy, *The Sanitarians: A History of American Public Health* (Urbana: University of Illinois Press, 1990), 122–23.

12. Offal removal accounted for $6,000 out of a budget of $17,056. Duffy, *The Sanitarians*, 143.

13. "Report of the Board of Public Improvements," *Mayor's Message with Accompanying Documents to the Municipal Assembly of the City of St. Louis at its Regular Session* (St. Louis: Times Printing House, 1878), 226, reel C19393, St. Louis Microfilm Collection, MSA.

14. Edward C. Rafferty, "The Boss Who Never Was: Colonel Ed Butler and the Limits of Practical Politics in St. Louis, 1875–1904," *Gateway Heritage* 12, no. 3 (Winter 1992): 59; Michael James Murray, "The Evolution of a City Boss: Ed Butler of St. Louis," (master's thesis, Northeast Missouri State University, 1975), 26.

15. "Annual Report of the Chief Sanitary Officer," *Fourteenth Annual Report of the Health Commissioner, City of St. Louis, MO, 1890–91*, 199, 201, MSA.

16. "Annual Report of the Health Commissioner," *Thirteenth Annual Report of the Health Commissioner, City of St. Louis, MO, 1889–1890*, 6, MSA.

17. "Annual Report of the Chief Sanitary Officer," *Thirteenth Annual Report*, 6.

18. See, e.g., George E. Waring, compiler, *Report on the Social Statistics of Cities Part II* (Washington, DC: Government Printing Office, 1887), 288.

19. Waring, *Report on the Social Statistics of Cities*, 286.

20. Hugh Miller Thompson, "Method Introduced by the Auxiliary Sanitary Association for Disposing of the Garbage of New Orleans," *Public Health Papers and Reports* 5 (1879): 33.

21. New Orleans's concealment of the outbreak led to its further spread across the South. See John H. Ellis, *Yellow Fever and Public Health in the New South* (Lexington: University Press of Kentucky, 1992); Margaret Warner, "Local Control versus National Interest: The Debate over Southern Public Health, 1878–1884," *Journal of Southern History* 50, no. 3 (August 1984): 407–28; Jerrold M. Michael, "The National Board of Health: 1879–1883," *Public Health Reports* 126, no. 1 (2011): 123–29.

22. Ellis, *Yellow Fever*, 354; Waring, *Report on the Social Statistics of Cities*, 286–87.

23. "The Widows' Carts," *Times-Democrat* (New Orleans), May 7, 1888, 4.

24. Carts may have been owned by ward bosses: "There can be no petty official steal in the steel garbage carts. The famous widows' carts put on the rolls by the ward boses [*sic*] who own them are not in the new deal, and, of course, the widows are in the dumps." "Our Picayunes," *Daily Picayune* (New Orleans), March 16, 1894, 4.

25. City of New Orleans, "1900 City Records: Carts," 17, NOPL.

26. Mrs. Olivia Babad to Tom Moulin, June 20, 1900, City of New Orleans, "1900 City Records: Carts," inserted at 16–17, NOPL.

27. City of St. Louis, Ordinance 15718, in *Ordinances, 1889–1890*, vol. 25, MSA.

28. "Annual Statement of the Health Department," *Seventeenth Annual Report of the Health Commissioner, City of St. Louis, for the year ending March 31, 1894*, 237, MSA.

29. "Annual Statement of the Health Department," *Seventeenth Annual Report*, 40.

30. "Annual Statement of the Health Department," *Seventeenth Annual Report*, 40.

31. "Annual Statement of the Health Department," *Nineteenth Annual Report of the Health Commissioner, City of St. Louis, for the year ending March 31, 1896*, 32, MSA.

32. "Annual Statement of the Health Department," *Nineteenth Annual Report*, 148, 31–32.

33. "Annual Report of the Clerk of Health Commissioner and Public Health," and "Report of the Health Commissioner," *Nineteenth Annual Report of the Health Commissioner*, 57, 32, MSA. City officials eventually decided that they would pay to dispose of household garbage but not commercial garbage. Commercial establishments could make their own contracts with the Sanitary Company for its disposal. "Annual Report of the Health Department," *The Mayor's Message with Accompanying Documents to the Municipal Assembly of the City of St. Louis for the Fiscal Year Ending April 12, 1897* (St. Louis: Nixon-Jones Printing Co., 1898), 179–82, MSA. The harbor and wharf commissioner fretted that these establishments would just throw their garbage in the river once again.

34. "Annual Statement of the Office of Health Commissioner and Board of Health," *Twenty-Fourth Annual Report of the Health Commissioner, City of St. Louis, for the Year Ending March 31, 1901*, 239, MSA.

35. "Annual Statement of the Office of Health Commissioner and Board of Health," *Twenty-Fourth Annual Report*, 239, 251–52.

36. "To Dispose of Garbage," *Daily Picayune* (New Orleans), December 10, 1888, 4.

37. "City Hall: Mr. Keppler's Garbage Ordinance Discussed in Committee," *Daily Picayune*, October 11, 1889, 8; "Handling Garbage," *Daily Picayune*, October 15, 1889, 4.

38. Brian Gary Ettinger, "John Fitzpatrick and the Limits of Working-Class Politics in New Orleans, 1892–1896," *Louisiana History* 26, no. 4 (Autumn 1985): 342.

39. "History of Major Burke's Frauds," *New York Times*, March 22, 1894, 4.

40. "Mayor Shakespeare [*sic*] Firm," *New York Times*, November 22, 1891, 1.

41. Miller Thompson, "Method Introduced," 33–34.

42. "Garbage Removal," *Times-Democrat*, April 13, 1893, 4.

43. "The Handling of Garbage" *Daily Picayune*, April 21, 1893, 4.

44. "The Garbage Question," *Daily Picayune*, May 30, 1893, 4.

45. "Garbage Removal," *Times-Democrat*, April 13, 1893, 4.

46. "Removal of Garbage—The Question Referred to the Board of Health," *Times-Democrat*, April 25, 1893, 4. According to some estimates, New Orleans

paid sixty-six cents per cubic yard to dispose of trash, less than New York (eighty-one cents) but more than Memphis (thirty-one cents). "Removal of Garbage," 4.

47. "The Handling of Garbage," *Daily Picayune*, April 21, 1893, 4.

48. "Garbage Disposal," *Daily Picayune*, April 21, 1893, 6.

49. "News Gathered at the City Hall," *Daily Picayune*, August 3, 1893, 12.

50. "The Garbage Job," *Daily Picayune*, August 15, 1893, 4.

51. Craig Colten, *An Unnatural Metropolis: Wresting New Orleans from Nature* (Baton Rouge: Louisiana State University Press, 2005), 59-60; William Francis Morse, *The Collection and Disposal of Municipal Waste* (New York: Municipal Journal and Engineer, 1908), 330.

52. "Municipal Affairs," *Daily Picayune*, February 3, 1894, 4.

53. Ettinger, "John Fitzpatrick," 344.

54. "The Garbage Business," *Daily Picayune*, August 21, 1893, 4.

55. "The Garbage Job," *Daily Picayune*, August 23, 1893, 4.

56. City of New Orleans, Committee on Public Health, Reports and Actions, November 9, 1892-November 2, 1896, 42-43, NOPL.

57. Joy Jackson, *New Orleans in the Gilded Age: Politics and Urban Progress, 1880-1896* (New Orleans: Louisiana Historical Association, 1987), 138.

58. Ettinger, "John Fitzpatrick," 355-56.

59. "Will the Garbage Contract Be Enforced?" *Daily Picayune*, July 26, 1895, 4.

60. "City Hall: Get Your Garbage Boxes," *Daily Picayune*, March 4, 1894, 10.

61. "That Garbage Contract," *Daily Picayune*, August 19, 1893, 4; "The Garbage Business," *Daily Picayune*, August 23, 1893, 4.

62. "The Garbage Job and Garbage Boxes," *Daily Picayune*, March 15, 1894, 4.

63. "The Garbage Job and Garbage Boxes," 4.

64. "The Infamous Garbage Ordinance," *Daily Picayune*, August 3, 1895, 4.

65. "Novel Methods Used to Force Collectors to Take Up Garbage," *St. Louis Republic*, September 4, 1902, 14.

66. McConachie, "The 'Big Cinch,'" 228-29.

67. Lawson, *American State Trials*, 9:492-93.

68. "Journal of the Council Regular Session," *St. Louis Republic*, May 16, 1901, 5.

69. City of St. Louis, *Twenty-Fifth Annual Report of the Health Commissioner for the Year Ending March 31, 1902*, 284, 285, MSA.

70. City of St. Louis, *Twenty-Fifth Annual Report*, 287, 278.

71. Lawson, *American State Trials*, 9:493-95.

72. Lawson, *American State Trials*, 9: 503-4; Steven L. Piott, *Holy Joe: Joseph W. Folk and the Missouri Idea* (Columbia: University of Missouri Press, 1997), 55; Lincoln Steffens, *The Shame of the Cities* (Mineola, NY: Dover Publications, 2012).

73. Lawson, *American State Trials*, 9:566.

74. Piott, *Holy Joe*, 55, citing letter from Roosevelt to reformer Joseph Folk, December 12, 1903.

75. Civic Improvement League of Saint Louis, *Second Annual Report of the Civic Improvement League*, vol. 2 (St. Louis: Gottschalk Printing Co., 1903), 13.

76. "Annual Statement of the Clerk of the Health Commissioner and Board of Health," *Twenty-Seventh Annual Report of the Health Commissioner, City of St. Louis, for the Fiscal Year Ending March 31, 1904*, 183, 128, MSA.

77. "Annual Report of the Secretary of the Board of Public Improvements," *Mayor's Message with Accompanying Documents to the Municipal Assembly of the City of St. Louis for the Fiscal Year Ending April 11, 1904*, MSA.

78. "Annual Report of the Secretary of the Board of Public Improvements," *Mayor's Message (1904)*; "Annual Statement of the Clerk of the Health Commissioner and Board of Health," *Twenty-Seventh Annual Report*, 128.

79. "Annual Statement of the Clerk of the Health Commissioner and Board of Health," *Twenty-Seventh Annual Report*, 129.

80. "Says Incineration Is the Best Method," *St. Louis Republic*, August 26, 1903, 1.

81. City of St. Louis, *Twenty-Seventh Annual Report*, 137–140.

82. City of St. Louis, *Twenty-Seventh Annual Report*, 194–95.

83. Council Bill no. 253, *Journal of the City Council, Regular Session, 1903–1904*, March 30, 1904, 684–85, MSA

84. Civic Improvement League, *Second Annual Report of the Civic Improvement League*, St. Louis, March 1904, vol 2, 13.

85. "Impeachment," *Times-Democrat*, July 28, 1894, 3.

86. "Will the Garbage Contract Be Enforced?" *Daily-Picayune*, July 26, 1895, 4.

87. "The City Council," *Times-Democrat*, August 28, 1895, 3.

88. "The Infamous Garbage Ordinance," *Daily Picayune*, August 3, 1895, 4.

89. "The Garbage Outrage," *Daily Picayune*, October 10, 1895, 4.

90. "The Infamous Garbage Ordinance."

91. "Councilman Maille's Fight against the Garbage Outrage," *Daily Picayune*, September 4, 1895, 4.

92. "Maurice Hart Indicted," *Daily Picayune*, August 14, 1895, 1.

93. Colten, *An Unnatural Metropolis*, 59–60; Morse, *The Collection and Disposal of Municipal Waste*, 6.

94. Haas, "John Fitzpatrick," 13.

95. Cited in Haas, "John Fitzpatrick," 7.

96. "Notes on Municipal Government," *Annals of the American Academy of Political and Social Science* 9, no. 1 (1897): 157.

97. "Discovery of 'Phillipine Island' in Mississippi River by Hiram Phillips Probably Gives to St. Louis Most Remarkable Reduction Plant in Country," *St. Louis Republic*, December 11, 1904, 31.

98. "Discovery of 'Phillipine Island.'"

99. "Discovery of 'Phillipine Island.'"

100. "Annual Report of the Street Commissioner," *Mayor's Message with Accompanying Documents to the Municipal Assembly of the City of St. Louis for the Fiscal Year Ending April 10, 1905*, 64–65, MSA.

101. Ord. no. 23526, approved September 30, 1907, *Mayor's Message with Accompanying Documents to the Municipal Assembly of the City of St. Louis for the*

Fiscal Year Ending April 12, 1909, November 26, 1909, 143, retrieved from Google Books, https://books.google.com.pr/books?id=xkMVAQAAMAAJ&lr=.

102. "Annual Report of the Street Commissioner," *Mayor's Message with Accompanying Documents to the Municipal Assembly of the City of St. Louis for the Fiscal Year Ending April 11, 1910,* 44, MSA.

103. "St. Louis Garbage Reduction Plant," *Municipal Journal and Engineer* 28, no. 14 (January–June 1910): 620.

104. "Street Cleaning and Refuse Disposal," *Municipal Journal* 34, no. 2 (February 6, 1913): 219; "Will Dispose of Garbage in River," *Municipal Journal* 34, no. 25 (June 19, 1913): 866.

105. "Will Dispose of Garbage in River," 866.

106. Haas, "John Fitzpatrick," 15, 16, 21.

107. Haas, "John Fitzpatrick," 21, 23.

108. Robert Williams, "Martin Behrman and New Orleans Civic Development, 1904–1920," *Louisiana History: The Journal of the Louisiana Historical Association* 2, no. 4 (Autumn 1961): 373–74.

109. T. J. Moulin, Commissioner, Department of Public Works, "Carts and Drivers: Issued to Foremen," 1901, Letters and Orders of the General Superintendent, Department of Public Works, NOPL.

110. "City Public Works: Commissioner Moulin's Interesting Review for the Year," unidentified newspaper clipping, Letters and Orders of the General Superintendent, Department of Public Works, NOPL.

111. George Smith to James A. Gleason, General Superintendent, January 4, 1908, Letters and Orders of the General Superintendent, Department of Public Works, NOPL.

112. Walter L. Dodd, *Report on the Health and Sanitary Survey of the City of New Orleans, 1918–1919* (New Orleans, 1919), 112, cited in Williams, "Martin Behrman," 394.

113. H. M. Mayo, Secretary, New Orleans Progressive Union, to Mayor and City Council, May 9, 1905; New Orleans Progressive Union to Mayor Martin Behrman, January 7, 1907, Mayor Martin Behrman Records, ser. 1, 1904–1920, NOPL.

114. George Smith to the President and Members of City Council, August 10, 1905, Letters and Orders of the General Superintendent: Department of Public Works, NOPL.

115. George Smith to the President and Members of City Council, October 17, 1905, Letters and Orders of the General Superintendent: Department of Public Works, NOPL.

116. George Smith to the President and Members of City Council, January 9, 1907, Letters and Orders of the General Superintendent: Department of Public Works, NOPL.

117. George Smith and J. A. Gleason to District Superintendents, April 10, 1907; letter from George Smith to L.G. Fitzgerald, Foreman, April 14, 1908; George Smith to Mr. Egan, Acting General Superintendent, May 12, 1910; George Smith to Joseph Gleason, January 22, 1909; and Joseph Gleason to All Superintendents, May 26, 1911, all in Letters and Orders of the General Superintendent: Department of Public Works, NOPL.

118. Williams, "Martin Behrman," 394–95.

4. Picking Up Trash

1. Municipal Record: Minutes of the Proceedings of the Council of the City of Pittsburgh for the Year 1875 (Pittsburgh: Herald Printing Company, 1876), 199, HSWP.

2. "The Health of Charleston—NO. 2," *Charleston Daily News*, March 23, 1869, 4.

3. James C. Scott, "Corruption, Machine Politics, and Political Change," *American Political Science Review* 63, no. 4 (December 1969): 1144.

4. James Banner, "The Problem of South Carolina," in *The Hofstadter Aegis: A Memorial*, ed. Stanley Elkins and Eric McKitrick (New York: Alfred A. Knopf, 1974), 60–93; Frederic Cople Jaher, *The Urban Establishment: Upper Strata in Boston, New York, Charleston, Chicago, and Los Angeles* (Urbana: University of Illinois Press, 1982), 365.

5. Joseph S. Nye, "Corruption and Political Development: A Cost-Benefit Analysis," *American Political Science Review* 61, no. 2 (June 1967): 419.

6. James Haw, "'The Problem of South Carolina' Reexamined: A Review Essay," *South Carolina Historical Magazine* 107, no. 1 (January 2006): 9–25; Lacy Ford, *Origins of Southern Radicalism: The South Carolina Upcountry, 1800–1860* (New York: Oxford University Press, 1988); Jaher, *The Urban Establishment*, 365.

7. Tom W. Shick and Don H. Doyle, "The South Carolina Phosphate Boom and the Stillbirth of the New South, 1867–1920," *South Carolina Historical Magazine*, 86, no. 1 (January 1985): 1–31.

8. Don H. Doyle, "Leadership and Decline in Postwar Charleston, 1865–1910," in *From the Old South to the New: Essays on the Transitional South*, ed. Walter J. Fraser and Winfred Moore (Westport, CT: Greenwood Press, 1981), 93–106.

9. Doyle, "Leadership and Decline."

10. Shick and Doyle, "The South Carolina Phosphate Boom," 4; Doyle, "Leadership and Decline," 94.

11. Layon Wayne Jordan, "Police and Politics: Charleston in the Gilded Age, 1880–1900," *South Carolina Historical Magazine* 80, no. 1 (January 1980): 35–50.

12. Compiled from data in Chalmers G. Davidson, *The Last Foray: The South Carolina Planters of 1860: A Sociological Study* (Columbia: University of South Carolina Press, 1971), 170–267; Jaher, *The Urban Establishment*, 365.

13. Jaher, *The Urban Establishment*, 405.

14. Harlan Greene, Harry S. Hutchins Jr., and Brian E. Hutchins, *Slave Badges and the Slave-Hire System in Charleston, South Carolina, 1783–1865* (Jefferson, NC: McFarland, 2008), 127, 138; Bernard E. Powers, *Black Charlestonians: A Social History, 1822–1885* (Fayetteville: University of Arkansas Press, 1994).

15. Francis G. Couvares, *The Remaking of Pittsburgh: Class and Culture in an Industrializing City* (Albany: State University of New York Press, 1984), 64.

16. John F. Bauman and Edward K. Muller, *Before Renaissance: Planning in Pittsburgh, 1889–1943* (Pittsburgh: University of Pittsburgh Press, 2006), 21.

17. "Flinn Takes the Saddle for His Candidate—Diehl: Magee Starts for South," *Pittsburg Post*, November 21, 1898, 2.

18. "Statement of Receipts and Expenditures by the City Council of Charleston, from the 1st of September 1850 to 1st September 1851," CCPL.

19. "Statement of Receipts"; "Cemeteries in Charleston, South Carolina," Find a Grave, 2022, https://www.findagrave.com/cemetery-browse/USA/South-Carolina/Charleston-County?id=county_2322.

20. Bauman and Muller, *Before Renaissance*, 21–24.

21. Lincoln Steffens, *The Shame of the Cities* (1904; repr., New York: Sagamore Press, 1957), 165.

22. Philip S. Klein and Ari Hoogenboom, *A History of Pennsylvania* (University Park: Pennsylvania State University Press, 1980).

23. *Thomas Loughrey et al. v. City of Pittsburgh, Henry I. Gourley, Controller of the City of Pittsburgh, and Booth & Flinn, Limited,* Court of Common Pleas, no. 1, Allegheny County (March Term, 1898), Scrapbooks and Letterbooks of William Flinn, 1898–1921, ASC.

24. By the mid-nineteenth century, Americans turned to ward politicians or even private groups or individuals for such vital urban services as water supply, street sanitation, and fire protection. Stanley K. Schultz and Clay McShane, "To Engineer the Metropolis: Sewers, Sanitation, and City Planning in Late Nineteenth-Century America," *Journal of American History* 65, no. 2 (September 1978): 389–411; "Letting Plans Made and Contracts Awarded," in *Annual Report of the Department of Public Works of the City of Pittsburgh for the Fiscal Year 1897* (Pittsburgh: Herald, 1898), 54–59, HSWP. For a description of underground utilities, see George A. Soper, *Modern Methods of Street Cleaning* (New York: Engineering News Publishing Co., 1909), 5–6.

25. James C. Scott, "Handling Historical Comparisons Cross-Nationally," in *Political Corruption: Concepts and Contexts,* ed. Arnold J. Heidenheimer and Michael Johnston (New Brunswick, NJ: Transaction Books, 2002), 132.

26. City of Charleston, *Digest of the Ordinances of the City Council of Charleston, from the year 1783 to July 1818* (Charleston: Archibald E. Miller, 1818), 222–24, HTDL.

27. City of Charleston, Journal of the Commissioner of Streets and Lamps, November 7, 1806, (1806–1818), CCPL; Christina Rae Butler, *Lowcountry at High Tide: A History of Flooding, Drainage, and Reclamation in Charleston, South Carolina* (Columbia: University of South Carolina Press, 2020).

28. City of Charleston, *Digest of the Ordinances, 1818,* 223.

29. City of Charleston, *Digest of Ordinances, 1818,* 49, 223.

30. City of Charleston, Journal of the Commissioner of Streets and Lamps, August 6, 1807, (1806–1818), CCPL.

31. John Horsey, comp., *Ordinances of the City of Charleston from the 14th September, 1854, to the 1st December, 1859* (Charleston: Walker, Evans & Co., 1859), 61–62, https://books.google.com.pr/books?id=ntVMAQAAMAAJ&printsec=frontcover&source=gbs_ge_summary_r&cad=0#v=onepage&q=Streets&f=false.

32. "Public Notice," *Charleston Daily News,* November 17, 1865, 3.

33. "Regular Meeting of City Council," *Charleston Daily News,* July 17, 1867, 4.

34. "Lazy Man's Load," *Charleston Daily News,* November 2, 1866, 5.

35. "Proceedings of City Council," *Charleston Daily News,* May 11, 1868, 4.

36. "Regular Meeting of Council," *Charleston Daily News,* April 25, 1868, 4.

37. "City Council—Council Chamber, July 9, 1868," *Charleston Daily News,* July 13, 1868, 4.

38. James M. McPherson, "Abolitionists and the Civil Rights Act of 1875," *Journal of American History* 52, no. 3 (December 1965): 497; John Oldfield, "On the Beat: Black Policemen in Charleston, 1869–1921," *South Carolina Historical Magazine* 102 (April 2001): 155.

39. G. Pillsbury to Major General O. O. Howard, January 10, 1870, in *Letters from the South, Relating to the Condition of Freedmen addressed to Major General O. O. Howard,* ed. J. W. Alvord (Washington, DC: Howard University Press, 1870), 10–11.

40. In another news article, Jenks is referred to as "the son-in-law of our Massachusetts mayor." "How the People's Money Goes," *Charleston Daily News,* July 31, 1871, 2; "Meeting of Council Last Night," *Charleston Daily News,* July 7, 1869, 3: "Joseph H. Jenks was declared elected. [J. H. Jenks is a stepson of the Mayor]." "Bonds and Securities of City Officers," *Charleston Daily News,* September 2, 1869, 3.

41. John Oldfield, "On the Beat: Black Policemen in Charleston, 1869–1921," *South Carolina Historical Magazine* 102, no. 2 (April 2001): 153–68; Walter J. Fraser Jr., *Charleston! Charleston! The History of a Southern City* (Columbia: University of South Carolina Press, 1989).

42. "City Inspector's Report," in *Annual Reports of the Officers of the City Government to the City Council of Charleston for the Year Ending December 31st, 1870* (Charleston: Republican Book and Job Office, 1871), 53–54, https://books. google.com/books?id=EwwtAQAAMAAJ&pg=PP5#v=onepage&q&f=false.

43. "Derelict Scavengers," *Charleston Daily News,* August 12, 1871, 3.

44. "How the People's Money Goes," *Charleston Daily News,* July 31, 1871, 2.

45. "An Invitation to Pestilence," *Charleston Daily News,* August 21, 1871, 3; "Meeting of Volunteer Assistants to Board of Health, Ward No. 4," *Charleston Daily News,* August 30, 1871, 3.

46. Letter from Secretary of War William Belknap to US Senate, "Communicating, In obedience to law, information in relation to quarantine on the Southern and Gulf coasts," December 6, 1872, 23, Senate Ex. Doc. no. 9, 42nd Cong., 3rd sess.

47. "Local Laconics," *Charleston Daily News,* December 3, 1872, 1.

48. "The City Fathers," *Charleston Daily News,* January 11, 1873, 4.

49. "Here the new House of Correction received 'vagrants and offenders of City Ordinances' who were required to labor during their terms of incarceration. The prisoners cultivated vegetables on the City Farm and maintained the grounds of the Public Cemetery." "Records of the Charleston House of Correction, 1868–1885," ed. Nicholas Michael Butler (2010), iii, CCPL.

50. "The City Council: Proceedings of the Regular Meeting Last Night," *Charleston Daily News,* February 19, 1873, 4. In this meeting one of the aldermen questioned the mayor regarding a $213,000 deficit found in the treasurer's books.

51. Fraser, *Charleston!* 297.

52. Governor Tillman's amendments to South Carolina's constitution established that "in order to vote persons were now required to pay a poll tax, a property tax, and demonstrate an understanding of the state's constitution, which also prohibited interracial marriage." Fraser, *Charleston!* 329.

53. Fraser, *Charleston!* 310.

54. Daniel J. Crooks and Douglas W. Bostick, *Charleston's Trial: Jim Crow Justice* (Charleston: The History Press, 2008), 24; Fraser, *Charleston!* 312.

55. Fraser, *Charleston!* 311.

56. Fraser, *Charleston!* 303, 308, 307, 310.

57. Christina Shedlock, "'Prejudicial to the Public Health': Class, Race, and the History of Land Reclamation, Drainage, and Topographic Alternation in Charleston, South Carolina, 1836–1940" (master's thesis, Graduate School of the College of Charleston and the Citadel, 2010).

58. "Report of Department of Health," *Charleston Year Book, 1895,* 84, CCPL.

59. John Duffy, "Hogs, Dogs, and Dirt: Public Health in Early Pittsburgh," *Pennsylvania Magazine of History and Biography* 87, no. 3 (July 1963): 297.

60. *A Digest of the Acts of Assembly, the Codified Ordinance of the City of Pittsburgh Adopted October 6, 1859* (Pittsburgh: W. S. Haven, 1860), 4–10, 119–20, https://www.google.com/books/edition/A_Digest_of_the_Acts_of_Assembly_the_Cod/qGhIAAAAYAAJ?hl=en; *A Digest of the Acts of Assembly relating to the General Ordinances of the City of Pittsburgh from 1804 to September 1, 1886* (Harrisburg: Edwin K. Meyers, 1887), 40–41.

61. "Sanitary Affairs," *Pittsburgh Daily Post,* April 3, 1880, 4; "The South Side Scourge," *Pittsburgh Post,* April 5, 1880, 4; "Sanitarians' Council," *Pittsburgh Daily Post,* May 1, 1880, 4.

62. Bauman and Muller, *Before Renaissance,* 20–24.

63. City of Pittsburgh, *Eighth Annual Report of the Department of Public Safety, 1895,* 297–313, HSWP.

64. "Questions of Sanitation: They Will Be Discussed by the State Associated Health Authorities," *Pittsburgh Post,* January 3, 1895, 8.

65. Pennsylvania Public Law 350, no. 258, sec. 20, June 26, 1895, in *A Digest of the Acts of Assembly relating to the General Ordinances of the City of Pittsburgh from 1804 to Nov. 12, 1908,* HSWP.

66. City of Pittsburgh, *Eighth Annual Report of the Department of Public Safety, 1895,* 313, 764–65.

67. Department of Public Safety, "Report of the Director, Contracts Awarded by the Director of the Department of Public Safety during the Fiscal Year Ending Jan. 31, 1900," in *Twelfth Annual Report of the Department of Public Safety, City of Pittsburgh, 1899* (Pittsburgh: Herald Publishing, 1900), 22, HSWP.

68. Irwin Osborn, "Disposal of Garbage by the Reduction Method," *American Journal of Public Health* 2, no. 12 (December 1912): 937.

69. Records of the Pittsburgh Select and Common Council, 1806–1938, box 7, folder 3, Standing Committees—Finance Committee, September 1910, HSWP; *Fisher et al v. American Reduction Co.,* 42 Atl. 36 (1899): 38–39.

70. Paul Hansen, "City Wastes Disposal and Street Cleaning," in *Proceedings of the Ohio Engineering Society, Twenty-Ninth Annual Meeting* (Columbus: Trauger Printing, 1908), 110–11.

71. Rudolph Hering and Samuel A. Greeley, *Collection and Disposal of Municipal Refuse* (New York: McGraw-Hill, 1921), 2.

72. M. N. Baker, ed. *Municipal Year Book* (New York: Engineering News Publishing Co., 1902), xix.

73. William Flinn to A. W. Dow, Esq., March 7, 1898, Scrapbooks and Letterbooks of William Flinn, 1898–1921, ASC.

74. *Fisher et al. v. American Reduction Co.*

75. William Flinn to John Fredericks, April 21, 1898; George Flinn to William Flinn, April 23, 1898, both in Scrapbooks and Letterbooks of William Flinn, 1898–1921, ASC.

76. William Flinn to F. G. Lauer, January 13, 1898; William Flinn to F. C. Lauer, February 2, 1898; William Flinn to F. C. Lauer, February 12, 1898; William Flinn to A. W. Dow, March 1, 1898; William Flinn to A. W. Dow, March 7, 1898, all in Scrapbooks and Letterbooks of William Flinn, 1898–1921, ASC.

77. William Flinn to John Fredericks, May 27, 1898, Scrapbooks and Letterbooks of William Flinn, 1898–1921, ASC.

78. William Flinn to T. F. Van Kirk, March 5, 1898, Scrapbooks and Letterbooks of William Flinn, 1898–1921, ASC.

79. William Flinn to Mr. C. Gray, January 31, 1898, Scrapbooks and Letterbooks of William Flinn, 1898–1921, ASC.

80. New York Bureau of Municipal Research, *The City of Pittsburgh, Pennsylvania: Report on a Survey of the Department of Public Health, et al.* (Pittsburgh: Pittsburgh Printing Co., 1913), 30.

81. Bauman and Muller, *Before Renaissance,* 40.

82. Civic Club of Allegheny County, "The Index: Celebrated Twentieth Anniversary Has Accomplished Much," Clippings—Civic Club History, 1902–1915, Civic Club of Allegheny County Records, 1896–1974, AIS.1970.02, ASC.

83. Civic Club of Allegheny County, "Sanitary Complaints" (1903), Civic Club of Allegheny County Records, 1896–1974, AIS.1970.02, ASC.

84. City of Pittsburgh, *Digest of the General Ordinances and Laws of the City of Pittsburgh to March 1, 1938,* 666, HSWP; New York Bureau of Municipal Research, *The City of Pittsburgh,* 23–24.

85. "Report of Department of Health," *Charleston Year Book, 1887,* 64; "Report of Department of Health," *Year Book, 1888,* 39; "Report of Department of Health," *Year Book, 1894,* 89, 124; "Report of Department of Health," *Year Book, 1903,* 77, CCPL.

86. E. T. Hiller, "Development of the Systems of Control of Convict Labor in the United States," *Journal of the American Institute of Criminal Law and Criminology* 5, no. 2 (1914): 241–69; Alex Lichtenstein, *Twice the Work of Free Labor: The Political Economy of Convict Labor in the New South* (New York: Verso, 1996), xvii, 7.

87. Fraser, *Charleston!* 323–24; "Report of Street Department," *Charleston Year Book, 1894,* 74–75; "Acts of the General Assembly of South Carolina," *Year Book, 1885,* 236–37; "Mayor Ficken's Annual Review," *Year Book, 1892,* 9; "Mayor Ficken's Annual Review," *Year Book, 1893,* 12; "Report of Street Department," *Year Book, 1894,* 74–75; "Report of the Board of Commissioners for the Management, Care & Custody of Convicts," *Year Book, 1896,* 297.

88. "Report of the Board of Commissioners for the Management, Care and Custody of Convicts," *Charleston Year Book, 1896,* 297.

89. "Acts of the General Assembly, 1899," "Report of Commissioners of Management of Convicts," *Charleston Year Book, 1899,* 308, 201.

90. "Report of Civil Engineer of Drainage Commission," *Charleston Year Book, 1902,* 93–99.

91. "Report of Chief of Police," *Charleston Year Book, 1903,* 133.

92. "Report of the Board of Commissioners for the Management, Care and Custody of Convicts," *Charleston Year Book, 1903,* 187.

93. "Report of Chief of Police," *Charleston Year Book, 1907,* 102.

94. Fraser, *Charleston!* 354.

95. "Report of Street Department," *Charleston Year Book, 1915,* 123–24.

96. "Report of Street Department," *Charleston Year Book, 1917,* 214–16.

97. "Report of Street Department," *Charleston Year Book, 1918,* 123, 121.

98. "Mayor Grace's Annual Review," *Charleston Year Book, 1920,* xxi–xxiii.

99. "Report of Street Department," *Charleston Year Book, 1921,* 117.

100. Jessica Trounstine, *Political Monopolies in American Cities: The Rise and Fall of Bosses and Reformers* (Chicago: University of Chicago Press, 2008).

101. Chris Potter, "Who Was William Flinn?" *Pittsburgh City Paper,* March 4, 2004, https://www.pghcitypaper.com/pittsburgh/in-his-autobiography-famous-muckraker-lincoln-steffens-mentions-that-pittsburgh-was-utterly-corrupt-worse-than-st-louis-and-that-william/Content?oid=1336081.

102. Philip Klein and Ari Hoogenboom, *A History of Pennsylvania* (University Park: Pennsylvania State University Press, 1980), 419–20.

103. "Grenet Is Given New Position," *Pittsburgh Press,* December 13, 1907, 1.

104. Klein and Hoogenboom, *History of Pennsylvania,* 421.

105. "Department of Public Health: Annual Report of the Director," *Annual Reports of the Executive Departments of the City of Pittsburgh for the Year Ending January 31, 1910,* vol. 1 (Pittsburgh: Murdoch-Kerr, 1910), 983–84, HTDL.

106. "Department of Public Health: Chief Clerk's Report," in *Annual Report of Mayor for Year Ending January 31, 1912* (Pittsburgh: Pittsburgh Printing Co., 1912), 480, HSWP.

107. "Report of Bureau of Sanitation," in *Annual Reports of Departments and Offices of the City of Pittsburgh, 1914,* 287, HSWP.

108. "Health Director Names Committees: Will Study Municipal Waste Disposal and Squatter Control," *Pittsburgh Post-Gazette,* October 2, 1933, 17; Mrs. R. Templeton Smith, "Cincinnati vs. Pittsburgh: Ohio City Saves Money in Nearly Every Phase of Government," *Pittsburgh Press,* March 9, 1933, 30.

109. "Herron–Flinn Real Estate Deal Made Mayoralty Issue," *Pittsburgh Press,* October 30, 1933, 4; "City Denies Payment of Refuse Bill: Complaints of Faulty Collection Aired in Council," *Pittsburgh Post-Gazette,* January 31, 1940, 13.

110. "Garbage Firm 'Rescues' City," *Pittsburgh Press,* March 30, 1940, 3; "Another Bungle," *Pittsburgh Press,* March 31, 1940, 26.

111. "Praise Given by Roessing: Incinerator Is Doing Well, Says Director," *Pittsburgh Post-Gazette,* April 18, 1940, 13.

112. "Garbage Bids Hearing Today," *Pittsburgh Post-Gazette,* July 15, 1931, 13.

5. Solving the Garbage Can Problem

1. Edwin A. Fisher, "President's Annual Address," *Municipal Engineering* 23, no. 5 (November 1902): 332.

2. "The Prospects for Young Engineers—A Symposium," *Engineering News* 32, no. 1 (July 5, 1894): 5.

3. Fisher, "President's Annual Address," 333.

4. "Outline of the Proceedings," *Municipal Engineering* 11, no. 5 (November 1896): 279.

5. "Garbage Disposal and Apparatus Therefor," *Municipal Engineering* 21, no. 3 (September 1901): 152.

6. "Recent Inventions," *Municipal Engineering* 23, no. 1 (July 1902): 61; "Recent Inventions," *Municipal Engineering* 23, no. 3 (September 1902): 219.

7. "Improvement and Contracting News," *Municipal Engineering* 21, no. 5 (November 1901): 376.

8. "Modern Methods of Refuse and Garbage Disposal," *Municipal Engineering* 52, no. 6 (June 1917): 318.

9. See Sarah A. Moore, "Garbage Matters: Concepts in New Geographies of Waste," *Progress in Human Geography* 36, no. 6 (2012): 780–99.

10. Untitled article, *Engineering News* 34, no. 8 (July 18, 1895): 40–41.

11. "New Garbage System to Be Given Tryout Next Week," *Louisville Herald*, February 17, 1918, Garbage Clippings, Garbage Disposal and Incinerators, bk. 2, LFPL.

12. Charles V. Chapin, *Municipal Sanitation in the United States* (Providence: Snow & Farnham, 1901), 672, 670.

13. New York Bureau of Municipal Research, *City of Pittsburgh, Pennsylvania: Report on a Survey of the Department of Public Health, June–July 1913* (Pittsburgh: Pittsburgh Printing Co., 1913), 24, 30, HPGTC.

14. Chapin, *Municipal Sanitation*, 133.

15. James C. Scott, *Seeing Like a State: How Certain Schemes to Improve the Human Condition Have Failed* (New Haven: Yale University Press, 1998).

16. "A Lady's Plan: To Bring About a Better System of Garbage Gathering," *Daily Picayune* (New Orleans), March 10, 1894, 3. "The Garbage Box Business," *Daily Picayune* March 10, 1894, 4.

17. "The Garbage Job and Garbage Boxes," *Daily Picayune*, March 15, 1894, 4.

18. "To the Garbage Box," *Daily Picayune*, March 18, 1894, 4.

19. Kathleen S. Sullivan and Patricia Strach, "Statebuilding through Corruption: Graft and Trash in Pittsburgh and New Orleans," in *Statebuilding from the Margins*, ed. Carol Nackenoff and Julie Novkov (Philadelphia: University of Pennsylvania Press, 2014), 95–117.

20. Joy Jackson, *New Orleans in the Gilded Age: Politics and Urban Progress, 1880–1896* (Baton Rouge: Louisiana State University Press, 1969), 138.

21. Michael Mann, *The Sources of Social Power*, vol. 3, *Global Empires and Revolution, 1890–1945* (New York: Cambridge University Press, 2012).

22. Michael Mann, "The Autonomous Power of the State: Its Origins, Mechanisms and Results," *European Journal of Sociology* 25, no. 2 (1984): 189, 190.

23. Mary McDowell, "Garbage as a Symbol of a City's Standards," Civic Club Papers, 1911–1919, Civic Club (Charleston, SC), Civic Club Records (1907–1955), SCHS.

24. Maureen A. Flanagan, "Gender and Urban Political Reform: The City Club and the Woman's City Club of Chicago in the Progressive Era," *American Historical Review* 95, no. 4 (October 1990): 1048; Martin V. Melosi, *Garbage in the Cities: Refuse, Reform, and the Environment* (Pittsburgh: University of Pittsburgh Press, 2005), 98; Suellen Hoy, *Chasing Dirt: The American Pursuit of Cleanliness* (New York: Oxford University Press, 1995).

25. *Good Housekeeping,* May 1917, 134; June 1917, 134; February 1917, 115; January 1917, 94; April 1917, 100, HTDL.

26. Mrs. F. Von A. Cabeen, "The Proper Disposal of Household Refuse," *American Kitchen Magazine: A Domestic Science Monthly* 3, no. 3 (June 1895): 97.

27. "A Kitchen Garbage Drier," *American Kitchen Magazine* 3, no. 3 (June 1895): 102.

28. *Good Housekeeping,* May 1917, 133; March 1917, 121, 146; June 1917, 136; December 1916, 158, HTDL.

29. Civic Club of Charleston, Civic Club Papers: Year Books, 1911–1919, Civic Club (Charleston, SC) Civic Club Records (1907–1955), SCHS.

30. Junior Civic League Report, December 16, 1914, Meeting Book, 1913–1917; Chairman, Committee on Ways and Means, to Mrs. Thomas Silcox, January 9, 1911, Civic Club Papers 1911–1919, Civic Club (Charleston, SC) Civic Club Records (1907–1955), SCHS.

31. Report, March 24, 1915, Meeting Book, 1913–1917, Civic Club (Charleston, SC) Civic Club Records (1907–1955), SCHS.

32. "The Tin Cans of the Junior Civic League," newspaper clipping dated April 5, 1912, Clippings, Civic Club (Charleston, SC) Civic Club Records (1907–1955), SCHS.

33. "Report of Street Department," *Charleston Year Book, 1920,* 126–127, CCPL.

34. Elizabeth Belanger, "The Neighborhood Ideal: Local Planning Practices in Progressive-Era Women's Clubs," *Journal of Planning History* 8, no. 2 (May 2009): 87–110.

35. "A League of Optimists," *Charities: A Weekly Review of Local and General Philanthropy* 11 (Week Ending September 12, 1903): 234.

36. "Compel the Collection of Garbage," *St. Louis Post-Dispatch,* August 30, 1902, 4.

37. "Ask for Women Sanitary Inspectors," *St. Louis Republic,* November 12, 1902, 5.

38. Civic Improvement League of St. Louis, *A Year of Civic Effort* (St. Louis: Civic League of St. Louis, 1907), 36; Civic League of St. Louis, *Year Book* (Chicago: Chicago School of Civics and Philanthropy, 1909), 39–40.

39. "St. Louis Garbage Reduction Plant," *Municipal Journal and Engineer* 28, no. 14 (April 27, 1910): 620; Civic League of St. Louis, *Year Book,* 40.

40. "Annual Report of the Health Department," *Twenty-Fourth Annual Report of the Health Commissioner, City of St. Louis, for the Year Ending March 31st, 1901* (St. Louis: Published by the Department, 1901), 238–40, 252, 257, MSA.

41. "Complain of Garbage Collection in West End," *St. Louis Republic,* October 28, 1900, 6.

42. W. J. Stevens, "Lessons of Three Cities in Municipal Adornment," *Brush and Pencil* 12, no. 6 (September 1903): 395–96.

43. Nancy T. Kinney, "The Impact of Restrictive Immigration Policies on Political Empowerment: Ethnic Organization Persistence in Early Twentieth Century St. Louis," *Citizenship Studies* 9, no. 1 (February 2005): 63–64; Michael Jones-Correa, "The Origins and Diffusion of Racial Restrictive Covenants," *Political Science Quarterly* 115, no. 4 (Winter 2000–2001): 548.

44. "Rear Yard Improvement," *St. Louis Republic,* July 1, 1903, 6.

45. "Ladies Discuss Sanitation," *St. Louis Republic,* November 19, 1903, 14.

46. "To Talk on Home Sanitation," *St. Louis Globe-Democrat,* October 6, 1904, 13.

47. "Council Committee Will Visit Ghetto to Study Sanitary Conditions," *St. Louis Post-Dispatch,* January 16, 1910, 33.

48. "Kreismann Takes Officials on a Garbage Junket," *St. Louis Post-Dispatch,* May 1, 1910, 30.

49. "If Anything Doesn't Suit You, Tell Civic League," *St. Louis Post-Dispatch,* December 9, 1911, 4.

50. *First Annual Report of the Civic Improvement League* (St. Louis: Gottschalk Printing Co., 1903), 9, 17–18.

51. Stevens, "Lessons of Three Cities," 395–96.

52. Angela Gugliotta, "Class, Gender, and Coal Smoke: Gender Ideology and Environmental Injustice in Pittsburgh, 1868–1914," *Environmental History* 5, no. 2 (April 2000): 174.

53. Civic Club of Allegheny County, *Fifteen Years of Civic History* (Pittsburgh: Nicholson Printing Co., 1910), 13–14, HPGTC.

54. "Pretty Hard Job Tackled: Civic Club Organized to Fight Prevalent Municipal Corruption Hard on the Local System," *Pittsburgh Daily Post,* October 8, 1895, 1.

55. Civic Club of Allegheny County Charter, Civic Club of Allegheny County Records, 1896–1974, AIS 1970:2, ASC.

56. *Fisher et al. v. American Reduction Co.*, 42 Atl. 36 (1899): 38–39.

57. Board of Directors, Minutes, 1895–1899, May 2, 1896, and November 5, 1897, Civic Club of Allegheny County Records, 1896–1974, AIS 1970:2, ASC.

58. Civic Club of Allegheny County, *Fifteen Years of Civic History,* 13–14.

59. Sanitary Complaints, 1903, Civic Club of Allegheny County Records, 1896–1974, AIS 1970:2, ASC.

60. "1915 List of Accomplishments," Civic Club of Allegheny County Records, 1896–1974, AIS 1970:2, ASC.

61. "Among the Clubs," *Pittsburgh Weekly Gazette,* October 16, 1904, 27.

62. "Colored Women's Clubs," *Pittsburgh Daily Post,* October 13, 1901, 4.

63. Laurence Glasco, ed., *The WPA History of the Negro in Pittsburgh* (Pittsburgh: University of Pittsburgh Press, 2004), 293–96.

64. Dorothy B. Porter, "The Organized Educational Activities of Negro Literary Societies, 1828–1846," *Journal of Negro Education* 5, no. 4 (October 1936): 555–76 (quotation at 557).

65. Glasco, *WPA History of the Negro in Pittsburgh*, 293.

66. Anne Firor Scott, "Most Invisible of All: Black Women's Voluntary Associations," *Journal of Southern History* 56, no. 1 (February 1990): 16.

67. Stephanie J. Shaw, "Black Club Women and the Creation of the National Association of Colored Women," *Journal of Women's History* 3, no. 2 (Fall 1991): 18.

68. Glasco, *WPA History of the Negro in Pittsburgh*, 296.

69. "An Appeal to Merchants," *Pittsburgh Press,* September 8, 1898, 5.

70. "A Strong Set of Resolutions," *Pittsburgh Courier,* August 26, 1911, 5.

71. *Minutes of the Eighth Biennial Session of the National Association of Colored Women at the Hampton Institute, 1912*, 39, Records of the National Association of Colored Women's Clubs, 1895–1992, pt. 1, Minutes of National Conventions, Publications, and President's Office Correspondence, Bethesda, MD, University Publications of America, 1994.

72. *Minutes of the Ninth Biennial Session of the National Association of Colored Women, 1914*, 39–40, Records of the National Association of Colored Women's Clubs, 1895–1992, pt. 1, Minutes of National Conventions, Publications, and President's Office Correspondence, Bethesda, MD, University Publications of America, 1994.

73. Helen Tucker, "The Negroes of Pittsburgh," in *Wage Earning Pittsburgh: The Pittsburgh Survey,* ed. Paul Kellogg (New York: Survey Associates, 1914), 426, HTDL.

74. Abraham Epstein, *The Negro Migrant in Pittsburgh* (Pittsburgh: University of Pittsburgh School of Economics, 1918), 12, 70.

75. *A History of the Club Movement Among the Colored Women of the United States of America, as contained in the Minutes of the Conventions, Held in Boston. . . . and of the National Federation of Afro-American Women, held in Washington . . . 1896*, 47, Records of the National Association of Colored Women's Clubs, 1895–1992, pt. 1, Minutes of National Conventions, Publications, and President's Office Correspondence, Bethesda, MD, University Publications of America, 1994.

76. Paul Underwood Kellogg, ed., *The Pittsburgh District: Civic Frontage. The Pittsburgh Survey* (New York: Survey Associates, 1914), 89–90, Russell Sage Foundation, https://www.russellsage.org/sites/all/files/Kellogg_The%20Pittsburgh%20District_0.pdf.

77. John Bauman and Edward Muller, *Before Renaissance: Planning in Pittsburgh, 1889–1943* (Pittsburgh: University of Pittsburgh Press, 2006), 39, 56.

78. General Committee on the Hill Survey, *Social Conditions of the Negro in the Hill District of Pittsburgh* (Pittsburgh: General Committee on the Hill Survey, 1930), 34, HPGTC.

79. New York Bureau of Municipal Research, *The City of Pittsburgh*, 24.

80. "Report of the Bureau of Sanitation," *Annual Reports of the Departments and Offices of the City of Pittsburgh, 1915*, 224. HSWP

81. "A Change of Heart," *Pittsburgh Gazette Times,* August 29, 1908, 4.

82. *A History of the Club Movement Among the Colored Women of the United States of America,* 37.

83. Chapin, *Municipal Sanitation*, 740.

84. "Refuse Collection and Disposal," *Municipal Journal and Engineer* 27, no. 8 (August 25, 1909): 301.

85. Rudolph Hering and Samuel Greeley, *Collection and Disposal of Municipal Refuse* (New York: McGraw-Hill, 1921), 244.

86. "Clean-Up Days Voted a Success," *Municipal Journal* 33, no. 24 (December 12, 1912): 876.

87. "Women Observe Waste Disposal," *Louisville Herald,* October 26, 1915, Garbage Clippings, Garbage Disposal and Incinerators, bk. 1, LFPL.

88. Mrs. Lee Bernheim, "A Campaign for Sanitary Collection and Disposal of Garbage," *The American City* 15, no. 2 (August 1916): 134–36; "Women Observe Waste Disposal," *Louisville Herald,* October 26, 1915.

89. "Chicago Social Worker Favors Incinerator for Garbage Disposal," *Louisville Courier-Journal,* October 27, 1915, Garbage Clippings, Garbage Disposal and Incinerators, bk. 1, LFPL.

90. "No Money for Garbage Plant," *Louisville Courier-Journal,* May 20, 1917, Garbage Clippings, Garbage Disposal and Incinerators, bk. 2, LFPL.

91. Bernheim, "Campaign."

92. Martin V. Melosi, *Garbage in the Cities: Refuse, Reform, and the Environment* (Pittsburgh: University of Pittsburgh Press, 2005), 101.

93. George C. Wright, "The NAACP and Residential Segregation in Louisville, Kentucky, 1914–1917," *Register of the Kentucky Historical Society* 78, no. 1 (Winter 1980): 39–40.

94. Tracy Campbell, "Machine Politics, Police Corruption, and the Persistence of Vote Fraud: The Case of Louisville, Kentucky, Election of 1905," *Journal of Policy History* 15, no. 3 (July 2003): 275.

95. John Kleber, *The Encyclopedia of Louisville* (Louisville: University Press of Kentucky, 2001), xxiv; *Buchanan v. Warley* 245 US 60 (1917).

96. Kleber, *Encyclopedia*, xxiii.

97. Jason Waterman and William Fowler, *Municipal Ordinances, Rules, Regulations Pertaining to Public Health, 1917–1919* (Washington, DC: Government Printing Office, 1921), 194.

98. "City to Experiment in Garbage Collection," *Louisville Courier-Journal,* February 15, 1918; "New Garbage System to Be Given Tryout Next Week," *Louisville Herald,* February 17, 1918, Garbage Clippings, Garbage Disposal and Incinerators, bk. 2, LFPL.

99. "Wet Garbage Collection in City Takes Big Slump," *Louisville Courier-Journal,* December 1, 1918, Garbage Clippings, Garbage Disposal and Incinerators, bk. 2, LFPL.

100. "City to Collect Garbage in the Point District," *Louisville Times,* August 7, 1920, Garbage Clippings, Garbage Disposal and Incinerators, bk. 3, LFPL.

101. "City Mechanizes Dirt War 'Troops,'" *Louisville Times,* July 23, 1940, Garbage Clippings, Louisville Street Cleaning Department, LFPL.

102. "City May Require Garbage Separated," *Louisville Courier-Journal,* August 2, 1941; "A Publicity Campaign for Garbage Disposal," *Louisville*

Courier-Journal, August 17, 1941; "Garbage Rule to Junk Pile: City Learns You Can't Buy Any Cans," *Louisville Times,* August 26, 1941, Garbage Clippings, Louisville Street Cleaning Department, LFPL.

103. "Mayor Asks Merchants to Help Keep Streets Clean: 3 Women Employed to Explain Garbage Separation," *Louisville Courier-Journal,* May 11, 1941; "Explains Garbage Plan: City Canvassers Going from House to House Get Few Complaints," *Louisville Times,* May 15, 1941; "Housewives Promise to Obey Garbage Rules," *Louisville Courier-Journal,* September 16, 1941, Louisville Street Cleaning Department, Clippings, LFPL; "Rats Become Louisville's No. 2 Nuisance: Filth and Garbage Bring Rats," *Louisville Courier-Journal,* March 23, 1941, Garbage Clippings, Garbage Disposal and Incinerators, bk. 5, LFPL.

104. M. N. Baker, ed., *Municipal Year Book, 1902* (New York: Engineering News Publishing Co., 1902), 143.

105. "In the Heart of the South: Letter to the Editor," *Medical Record* 38, no. 13 (September 27, 1890), 366.

106. Morris Knowles, "Water and Waste: The Sanitary Problems of a Modern Industrial District," *The Survey,* 27, no. 14 (January 6, 1912): 1499.

107. Charles Connerly, *"The Most Segregated City in America": City Planning and Civil Rights in Birmingham, 1920–1980* (Charlottesville: University of Virginia Press, 2005).

108. Roger Biles, *The South and the New Deal* (Lexington: University Press of Kentucky, 1994), 107.

109. Robert W. Widell, *Birmingham and the Long Black Freedom Struggle* (New York: Palgrave Macmillan, 2013).

110. *The Crisis* 13, no. 1 (November 1916): 23.

111. *The Crisis* 6, no. 1 (May 1913): 8; *The Crisis* 7, no. 2 (December 1913): 60; *The Crisis* 8, no. 2 (February 1914): 164.

112. Rheta Louise Childe Dorr, *What Eight Million Women Want* (Boston: Small, Maynard & Co., 1910), 38; "Eight Million Women," *La Follette's Weekly Magazine* 1, no. 34 (August 28, 1909): 10.

113. Carl V. Harris, "Reforms in Government Control of Negroes in Birmingham, Alabama, 1890–1920," *Journal of Southern History* 38, no. 4 (November 1972): 567–600; W. David Lewis, "The Emergence of Birmingham as a Case Study of Continuity between the Antebellum Planter Class and Industrialization in the 'New South,'" *Agricultural History* 68, no. 2 (Spring 1994): 62–79.

114. *The Survey* implied this was a medieval practice. See Knowles, "Water and Waste," 1499.

115. "When Birmingham Went Broke," *The Survey,* 34, no. 24 (September 11, 1915): 530; *Public Health Administration: City of Birmingham and County of Jefferson, Alabama* (Washington, DC: Government Printing Office, 1916), 21; W. David Lewis, *Sloss Furnaces and the Rise of the Birmingham District: An Industrial Epic* (Tuscaloosa: University of Alabama Press, 1994), 306.

116. Martha Hodes, "The Sexualization of Reconstruction Politics: White Women and Black Men in the South after the Civil War," *Journal of the History of Sexuality* 3, no. 3 (January 1993): 402–17; Barbara Welke, "When All the Women Were White, and All the Blacks Were Men: Gender, Class, Race,

and the Road to *Plessy*, 1855–1914," *Law & History Review* 13, no. 2 (Autumn 1995) 261–316.

117. "There's a Right Way to Do a Thing . . .," *Louisville Courier-Journal*, September 28, 1941, Garbage Clippings, Louisville Street Cleaning Department, LFPL.

118. "Garbage Campaign Flops, Early Survey Indicates," *Louisville Times*, September 2, 1941, Garbage Clippings, Louisville Street Cleaning Department, LFPL.

6. Getting and Keeping Garbage Collection

1. Julie Novkov, *Racial Union: Law, Intimacy, and the White State in Alabama, 1865–1954* (Ann Arbor: University of Michigan Press, 2008); Rogers Smith, "Beyond Tocqueville, Myrdal, and Hartz: The Multiple Traditions in America," *American Political Science Review* 87, no. 3 (September 1993): 549–66.

2. See Louie Davis Shivery and Hugh H. Smythe, "The Neighborhood Union: A Survey of the Beginnings of Social Welfare Movements among Negroes in Atlanta," *Phylon* 3, no. 2 (2nd quarter 1942): 149–62; Jacqueline A. Rouse, "The Legacy of Community Organizing: Lugenia Burns Hope and the Neighborhood Union," *Journal of Negro History* 69, no. 3–4 (Summer–Autumn 1984): 114–33.

3. Carl Zimring, "Dirty Work: How Hygiene and Xenophobia Marginalized the American Waste Trades, 1870–1930," *Environmental History* 9, no. 1 (January 2004): 84; Carl A. Zimring, *Clean and White: A History of Environmental Racism in the United States* (New York: New York University Press, 2015).

4. Michael K. Honey, *Going Down Jericho Road: The Memphis Strike, Martin Luther King's Last Campaign* (New York: W. W. Norton & Co., 2007), 418.

5. Fifty-six percent of the population according to the census in 1890 and 1900. Campbell Gibson and Kay Jung, "Historical Census Statistics on Population Totals by Race, 1790 to 1990, and by Hispanic Origin, 1970 to 1990, for Large Cities and Other Urban Places in the United States," US Census Bureau working paper, February 2005, https://www.census.gov/library/working-papers/2005/demo/POP-twps0076.html.

6. George A. Soper, *Modern Methods of Street Cleaning* (New York: Engineering News Publishing Co., 1909), 3.

7. William P. McGowan, "American Wasteland: A History of America's Garbage Industry: 1880–1989," *Business and Economic History* 24, no. 1 (Fall 1995): 157.

8. Soper, *Modern Methods of Street Cleaning*, 3, 4; Lawrence H. Larsen, "Nineteenth-Century Street Sanitation: A Study of Filth and Frustration," *Wisconsin Magazine of History* 52, no. 3 (Spring 1969): 242–43.

9. John Duffy, "Hogs, Dogs, and Dirt: Public Health in Early Pittsburgh," *Pennsylvania Magazine of History and Biography* 87, no. 3 (July 1963): 297.

10. *A Digest of the Ordinances of the City of Pittsburgh* (Pittsburgh: W. H. Whitney, 1849), 149, HPGTC.

11. George Fleming, *History of Pittsburgh and Its Environs*, vol. 2 (New York: American Historical Society, 1922), 67.

12. *A Digest of the Acts of Assembly, the Codified Ordinance of the City of Pittsburgh Adopted October 6, 1859* (Pittsburgh: W. S. Haven, 1860), 5, HGPTC.

13. *A Digest of the Acts of Assembly Relating to, and the General Ordinances of the City of Pittsburgh, from 1804 to September 1, 1886* (Harrisburg: Edwin Meyers, 1887), 445, https://www.google.com/books/edition/A_Digest_of_the_Acts_of_Assembly_Relatin/Rs5CAAAAYAAJ?hl=en&gbpv=0.

14. Larsen, "Nineteenth-Century Street Sanitation," 243.

15. *Digest of the Acts of Assembly, 1804–1886*, 416.

16. *Municipal Record: Minutes of the Proceedings of the Select and Common Councils of the City of Pittsburgh for the Year 1875* (Pittsburgh: Herald Printing Co., 1876), 199, HSWP.

17. "Pittsburgh Smoke: It Signifies Prosperity, Yet We Would Like to Get Rid of It," *Commercial Gazette,* October 3, 1879, 4.

18. "Fever Epidemic: South Siders Suffering from Typhoid Fever," *Pittsburgh Daily Post,* April 2, 1880, 4.

19. "Sanitary Affairs: South Side Fever Considered by the Board of Health," *Pittsburgh Daily Post,* April 3, 1880, 4.

20. "The Sanitarians: Dumping Garbage—Last Month's Mortality Record," *Pittsburgh Daily Post,* March 13, 1880, 4.

21. "Sanitarians' Council Talk Over the South Side Typhoid Fever Epidemic," *Pittsburgh Daily Post,* May 1, 1880, 4.

22. Carl A. Zimring, *Cash for Your Trash: Scrap Recycling in America* (New Brunswick: Rutgers University Press, 2009).

23. *Municipal Record: Minutes for 1875*, 199.

24. Carl Zimring, "Dirty Work."

25. "Rag Pickers: To What Extent Their Business Is Carried on in this City," *Pittsburgh Daily Post,* February 27, 1880, 4.

26. "South Side Health: Report of the Sanitarians on the Recent Fever Epidemic," *Pittsburgh Daily Post,* June 1, 1880, 4.

27. Angela Gugliotta, "Class, Gender, and Coal Smoke: Gender Ideology and Environmental Injustice in Pittsburgh, 1868-1914," *Environmental History* 5, no. 2 (April 2000): 170, 174.

28. "A Chapter on Junk-Shops." *Charleston Daily News,* February 18, 1868, 3; "The Charleston Chiffoniers," *Charleston Daily News,* September 20, 1867, 3.

29. "The Health of Charleston—NO. 2," *Charleston Daily News,* March 23, 1869, 4.

30. "Lazy Man's Load," *Charleston Daily News,* November 2, 1866, 5; "Bill for the Better Regulation of the Street Department," *Charleston Daily News,* May 11, 1868, 4.

31. "The Health of Charleston—NO. 2."

32. "Main Stationhouse, Office of Captain of Police, Charleston, S.C.," *Charleston Daily News,* August 30, 1871, 2.

33. "Report of Health Department," *Charleston Year Book, 1880,* 30; "Report of Department of Health," *Year Book, 1897,* 90-114, CCPL.

34. "Report of Department of Health," *Charleston Year Book, 1883,* 52, CCPL.

35. A 1917 *Pittsburgh Press* article mentions that the garbage collectors in nearby Allegheny were all Black. "Serious Situation May Be Caused by Garbage Strike," *Pittsburgh Press,* March 7, 1917, 6.

36. Helen Tucker, "The Negroes of Pittsburgh," in *Wage Earning Pittsburgh: The Pittsburgh Survey,* vol. 6, ed. Paul Kellogg (New York: Survey Associates, 1914), 429, 121–22 HTDL.

37. "Serious Situation May Be Caused by Garbage Strike," *Pittsburgh Press,* March 7, 1917, 6.

38. "He's Scarce Since Saloons Have Closed," *Pittsburgh Daily Post,* January 16, 1920, 5.

39. "Official—Pittsburgh," *Pittsburgh Press,* May 10, 1904, 20.

40. "Should Have Known Their Garbage Man Better," *Pittsburgh Press,* December 30, 1915, 13.

41. "Scarlet Fever Kills Three in One Family," *Pittsburgh Press,* May 7, 1902, 11.

42. "Report of the Bureau of Sanitation," in *Annual Reports of the Departments and Offices of the City of Pittsburgh, 1914* (Pittsburgh: Pittsburgh Printing Co., 1914), 287, HSWP.

43. "Mayor Ficken's Annual Review," *Charleston Year Book, 1892,* 7; "Report of Department of Health," *Year Book, 1888,* 57, CCPL.

44. City of Charleston, *Digest of the Ordinances of the City Council of Charleston, from the year 1783 to July 1818* (Charleston: Archibald E. Miller, 1818), 49, 223, HTDL.

45. "Report of Superintendent of Streets," *Charleston Year Book, 1880,* 169; "Report of Department of Health," *Year Book, 1887,* 59, CCPL.

46. "Report of Board of Public Works," *Charleston Year Book, 1907,* 67; "Report of Board of Public Works," *Year Book, 1908,* 57, CCPL.

47. "Report of Chief of Police," *Charleston Year Book, 1893,* 154–56, CCPL.

48. Matthew J. Mancini, "Race, Economics, and the Abandonment of Convict Leasing," *Journal of Negro History* 63, no. 4 (Fall 1978): 339–52; Steve Fraser and Joshua B. Freeman, "In the Rearview Mirror: Barbarism and Progress: The Story of Convict Labor," *New Labor Forum* 21, no. 3 (Fall 2012): 94–98.

49. "Report of Department of Health," *Charleston Year Book, 1903,* 69, 77, CCPL.

50. "Report of Department of Health," *Charleston Year Book, 1903,* 77; "Report of Garbage Collection Department," *Year Book, 1905,* 66; "Report of Board of Public Works," *Year Book, 1906,* 71, CCPL.

51. "Report of Department of Health," *Charleston Year Book, 1905,* 132, CCPL.

52. "Report of Department of Health," *Charleston Year Book, 1905,* 132; "Mayor Grace's Annual Report," *Year Book, 1912,* xvi, CCPL.

53. "Mayor Grace's Annual Report," *Charleston Year Book, 1912,* xvi–xvii, CCPL.

54. "Report of Department of Health," *Charleston Year Book, 1915,* 151, CCPL.

55. "Mayor Grace's Annual Report," *Charleston Year Book, 1914,* xvii, CCPL.

56. Junior Civic League Report, December 16, 1914, Civic Club Papers: 1911–1919, Civic Club (Charleston, SC), Civic Club Records, 1907–1955, SCHS.

57. "Report of Street Department," *Charleston Year Book, 1916*, 159; "Report of Department of Health," *Year Book, 1916*, 218, CCPL.

58. "Report of Street Department," *Charleston Year Book, 1917*, 213–16; "Report of Street Department," *Year Book, 1918*, 119; "Report of Street Department," *Year Book, 1920*, 121, CCPL.

59. Nayan Shah, *Contagious Divides: Epidemics and Race in San Francisco's Chinatown* (Berkeley: University of California Press, 2001), 49.

60. Philip Ethington, "Vigilantes and the Police: The Creation of a Professional Police Bureaucracy in San Francisco, 1847–1900," *Journal of Social History* 21, no. 2 (Winter 1987): 202–3.

61. R. A. Burchell, *The San Francisco Irish, 1848–1880* (Manchester: Manchester University Press, 1980), 27–28.

62. Shah, *Contagious Divides*, 50.

63. "Health Officer's Report," *San Francisco Municipal Reports for the Fiscal Year 1871–72* (San Francisco: Cosmopolitan Printing Co., 1872), 216, SFPL.

64. "Health Officer's Report," *San Francisco Municipal Reports for the Fiscal Year 1884–1885* (San Francisco: W. M. Hinton & Co., 1885), 228–29, SFPL.

65. "Farwell's Garbage Order," *Daily Evening Bulletin* (San Francisco), May 13, 1885, 3.

66. "The Scavenger's Union," *Daily Evening Bulletin*, October 20, 1886, 1.

67. "The Garbage Question," *Daily Evening Bulletin*, May 6, 1887, 2.

68. "A New Scheme to Circumvent the Scavengers," *San Francisco Call*, May 16, 1899, 10; "Garbage Contract," *San Francisco Call*, January 4, 1899, 9.

69. "Injunction Is Granted in the Garbage Suit," *San Francisco Call*, May 26, 1899, 6.

70. "Report of Board of Health," *San Francisco Municipal Reports, 1899–1900* (San Francisco: Hinton Printing Co., 1900), 545, SFPL.

71. William Stiles, "The San Francisco Earthquake and Fire—Public Health Aspects," *California Medicine* 85, no. 1 (July 1956): 36.

72. Dan Kurzman, *Disaster! The Great San Francisco Earthquake and Fire of 1906* (New York: Harper Perennial, 2002), 158.

73. W. C. Hassler, "Sanitation in San Francisco from April 18th to August 1st, 1906," *California State Journal of Medicine* 4, no. 9 (September 1906): 243.

74. Stiles, "The San Francisco Earthquake and Fire," 36.

75. Michael J. Sangiacomo, "Achieving Zero Waste in San Francisco: Successes and Challenges," *Journal of International Affairs* 73, no. 1 (Fall 2019/Winter 2020): 285–90.

76. Martin V. Melosi, *Garbage in the Cities: Refuse, Reform, and the Environment* (Pittsburgh: University of Pittsburgh Press, 2005), 173; Stewart E. Perry, *Collecting Garbage: Dirty Work, Clean Jobs, Proud People* (New Brunswick, NJ: Transaction Publishers, 1998), 18.

77. City Sanitation Guild, "Housecleaning," flyer dated February 1929, Garbage—City Sanitation Guild, 1929, San Francisco Ephemera Collection, SFPL.

78. City Sanitation Guild, "We Need Your Help," flyer dated March 1929, Garbage–City Sanitation Guild, 1929, San Francisco Ephemera Collection, SFPL.

79. City Sanitation Guild, "A Plain Talk to Housewives about the Garbage Situation," flyer dated May 1929, Garbage–City Sanitation Guild, 1929, San Francisco Ephemera Collection, SFPL, emphasis added.

80. City Sanitation Guild, "What You Owe Your Scavenger," undated flyer, Garbage–City Sanitation Guild, 1929, San Francisco Ephemera Collection, SFPL.

7. The Politics of Garbage Collection

1. Carol Nackenoff and Julie Novkov, "Statebuilding in the Progressive Era: A Continuing Dilemma in American Political Development," in *Statebuilding from the Margins: Between Reconstruction and the New Deal*, ed. Carol Nackenoff and Julie Novkov (Philadelphia: University of Pennsylvania Press, 2014), 1–31.

2. Theda Skocpol and Kenneth Finegold, "State Capacity and Economic Intervention in the Early New Deal," *Political Science Quarterly* 97, no. 2 (1982): 255–78.

3. Theda Skocpol, "Bringing the State Back In: Strategies of Analysis in Current Research," in *Bringing the State Back In*, ed. Peter B. Evans et al. (Cambridge: Cambridge University Press, 1985), 9.

4. Emphasis in the original. Luciana Cingolani, "The Role of State Capacity in Development Studies," *Journal of Development Perspectives* 2, no. 1–2 (2018): 106; Elissa Berwick and Fotini Christia, "State Capacity Redux: Integrating Classical and Experimental Contributions to an Enduring Debate," *Annual Review of Political Science* 21 (2018): 71–91.

5. Stephen Skowronek, *Building a New American State: The Expansion of National Administrative Capacities, 1877–1920* (New York: Cambridge University Press, 1982).

6. Daniel P. Carpenter, *The Forging of Bureaucratic Autonomy: Reputations, Networks, and Policy Innovation in Executive Agencies, 1862–1928* (Princeton: Princeton University Press, 2002), 28.

7. Patricia Strach and Kathleen Sullivan, "Dirty Politics: Public Employees, Private Contractors, and the Development of Nineteenth-Century Trash Collection in Pittsburgh and New Orleans," *Social Science History* 39, no. 3 (Fall 2015): 387–407.

8. Ruth Bloch Rubin, "State Preventative Medicine: Public Health, Indian Removal, and the Growth of State Capacity, 1800-1840," *Studies in American Political Development* 34, no. 1 (April 2020): 24–43; William Adler, *Engineering Expansion: The U.S. Army and Economic Development, 1787–1860* (Philadelphia: University of Pennsylvania Press, 2021); Andrew S. Kelly, "The Political Development of Scientific Capacity in the United States," *Studies in American Political Development* 28, no. 1 (2014): 1–25; Brian Balogh, *A Government Out of Sight: The Mystery of National Authority in Nineteenth-Century America* (New York: Cambridge University Press, 2009), 379.

9. William Novak, *The People's Welfare: Law and Regulation in Nineteenth-Century America* (Chapel Hill: University of North Carolina Press, 1996).

10. Daniel Sledge, *Health Divided: Public Health and Individual Medicine in the Making of the Modern American State* (Lawrence: University Press of Kansas, 2017).

11. Kimberley S. Johnson, "Modernity, Public Administration, and the Disappearance of the American States: A Necessary Development?" *Administration & Society* 35, no. 2 (May 2003): 144-59.

12. Colin D. Moore, "State Building through Partnership: Delegation, Public-Private Partnerships, and the Political Development of American Imperialism, 1898-1916," *Studies in American Political Development* 25, no. 1 (2011): 29.

13. Kelly, "The Political Development of Scientific Capacity," 1.

14. Theda Skocpol, *Protecting Soldiers and Mothers: The Political Origins of Social Policy in the United States* (Cambridge: Belknap Press of Harvard University Press, 1992); Carpenter, *The Forging of Bureaucratic Autonomy;* Daniel Carpenter, *Reputation and Power: Organizational Image and Pharmaceutical Regulation at the FDA* (Princeton: Princeton University Press, 2010); Daniel P. Carpenter, "State Building through Reputation Building: Coalitions of Esteem and Program Innovation in the National Postal System, 1883-1913," *Studies in American Political Development* 14, no. 2 (October 2000): 121-55.

15. Carpenter, *The Forging of Bureaucratic Autonomy;* Carpenter, *Reputation and Power;* Carpenter, "State Building through Reputation Building."

16. Christopher Howard, *The Hidden Welfare State: Tax Expenditures and Social Policy in the United States* (Princeton: Princeton University Press, 1997); Marie Gottschalk, *The Shadow Welfare State: Labor, Business, and the Politics of Health Care in the United States* (Ithaca: Cornell University Press, 2000); Suzanne Mettler, *The Submerged State: How Invisible Government Policies Undermine American Democracy* (Chicago: University of Chicago Press, 2011); Patricia Strach, *All in the Family: The Private Roots of American Public Policy* (Stanford: Stanford University Press, 2007); Patricia Strach, *Hiding Politics in Plain Sight: Cause Marketing, Corporate Influence, and Breast Cancer Policymaking* (New York: Oxford University Press, 2016); Kimberly J. Morgan and Andrea Louise Campbell, "Delegated Governance in the Affordable Care Act," *Journal of Health Politics, Policy and Law* 36, no. 3 (June 2011): 387-91.

17. Carol Nackenoff, "The Private Roots of American Political Development: The Immigrants' Protective League's 'Friendly and Sympathetic Touch,' 1908-1924," *Studies in American Political Development* 28, no. 2 (October 2014): 129.

18. Marek D. Steedman, "Demagogues and the Demon Drink: Newspapers and the Revival of Prohibition in Georgia"; Carol Nackenoff and Kathleen S. Sullivan, "The House That Julia (and Friends) Built: Networking Chicago's Juvenile Court"; and Ann-Marie Syzmanski, "Wildlife Protection and the Development of Centralized Governance in the Progressive Era," all in Nackenoff and Novkov, *Statebuilding from the Margins,* 65-94, 171-202, and 140-70.

19. James L. Greer, "The Better Homes Movement and the Origins of Mortgage Redlining in the United States," in Nackenoff and Novkov, *Statebuilding from the Margins,* 203-35.

20. James T. Sparrow, William J. Novak, and Stephen W. Sawyer, eds., *Boundaries of the State in US History* (Chicago: University of Chicago Press, 2015).

21. Jessica Wang, "Dogs and the Making of the American State: Voluntary Association, State Power, and the Politics of Animal Control in New York City, 1850–1920," *Journal of American History* 98, no. 4 (March 2012): 998–1024.

22. Gabriel N. Rosenberg, *The 4-H Harvest: Sexuality and the State in Rural America* (Philadelphia: University of Pennsylvania Press, 2015).

23. See Chloe N. Thurston, *At the Boundaries of Homeownership: Credit, Discrimination, and the American State* (New York: Cambridge University Press, 2018); Virginia Eubanks, "Technologies of Citizenship: Surveillance and Political Learning in the Welfare System," in *Surveillance and Security: Technological Politics and Power in Everyday Life*, ed. Torin Monahan (New York: Routledge, 2006), 89–108.

24. Carl Zimring, "Dirty Work: How Hygiene and Xenophobia Marginalized the American Waste Trades, 1870–1930," *Environmental History* 9, no. 1 (January 2004): 80–101; Carl A. Zimring, *Cash for Your Trash: Scrap Recycling in America* (New Brunswick: Rutgers University Press, 2009).

25. See Berwick and Christia, "State Capacity Redux."

26. Carpenter, *The Forging of Bureaucratic Autonomy*, 41; Skowronek, *Building a New American State*.

27. Carpenter, *The Forging of Bureaucratic Autonomy*, 34.

28. Carpenter, *The Forging of Bureaucratic Autonomy*, 47.

29. Moore, "State Building through Partnership," 34.

30. Desmond King and Robert C. Lieberman, "Finding the American State: Transcending the 'Statelessness' Account," *Polity* 40, no. 3 (July 2008): 15; Jamila Michener, *Fragmented Democracy: Medicaid, Federalism, and Unequal Politics* (New York: Cambridge University Press, 2018).

31. Michener, *Fragmented Democracy*; Suzanne Mettler, *Dividing Citizens: Gender and Federalism in New Deal Public Policy* (Ithaca: Cornell University Press, 1998); Robert Lieberman, *Shifting the Color Line: Race and the American Welfare State* (Cambridge: Harvard University Press, 2001).

32. Rogers M. Smith, *Civic Ideals: Conflicting Visions of Citizenship in U.S. History* (New Haven: Yale University Press, 1997); Desmond S. King and Rogers M. Smith, "Racial Orders in American Political Development," *American Political Science Review* 99, no. 1 (February 2005): 75–92.

33. Chloe Thurston, "Black Lives Matter: American Political Development and the Politics of Visibility," *Politics, Groups and Identities* 6 (2019): 166.

34. Joseph Lowndes, Julie Novkov, and Dorian Warren, eds., *Race and American Political Development* (New York: Routledge, 2008).

35. Kimberley S. Johnson, "The Color Line and the State: Race and American Political Development," in *The Oxford Handbook of American Political Development*, ed. Richard Valelly et al. (New York: Oxford University Press, 2016), 596.

36. Julie Novkov, "Bringing the States Back In: Understanding Legal Subordination and Identity through Political Development," *Polity* 40, no. 1 (January 2008): 24–48.

37. Karen Orren and Stephen Skowronek, *The Search for American Political Development* (Cambridge: Cambridge University Press, 2004).

38. Carol Nackenoff and Julie Novkov, "Statebuilding in the Progressive Era" 7.

39. Colin Gordon, *Citizen Brown: Race, Democracy, and Inequality in the St. Louis Suburbs* (Chicago: University of Chicago Press, 2019).

40. Theodore M. Shaw, *Investigation of the Ferguson Police Department* (Washington, DC: US Department of Justice, Civil Rights Division, 2015), https://www.justice.gov/sites/default/files/opa/press-releases/attachments/2015/03/04/ferguson_police_department_report.pdf.

41. Jessica Trounstine, *Segregation by Design: Local Politics and Inequality in American Cities* (New York: Cambridge University Press, 2018).

42. See Trounstine, *Segregation by Design*; Gordon, *Citizen Brown*.

43. Johnson, "Modernity, Public Administration, and the Disappearance of American States."

44. Kimberley S. Johnson, *Governing the American State: Congress and the New Federalism, 1877–1929* (Princeton: Princeton University Press, 2007).

45. Kimberley S. Johnson, *Reforming Jim Crow: Southern Politics and State in the Age before Brown* (New York: Oxford University Press, 2010), 4.

46. See Mettler, *Dividing Citizens*; Ange-Marie Hancock, *The Politics of Disgust: The Public Identity of the Welfare Queen* (New York: New York University Press, 2004); Amy E. Lerman and Vesla M. Weaver, *Arresting Citizenship: The Democratic Consequences of American Crime Control* (Chicago: University of Chicago Press, 2014); Vesla M. Weaver, "Frontlash: Race and the Development of Punitive Crime Policy," *Studies in American Political Development* 21, no. 2 (Fall 2007): 230–65.

47. See Richardson Dilworth, "Introduction: Bringing the City Back In," and Clarence N. Stone and Robert K. Whelan, "Through a Glass Darkly: The Once and Future Study of Urban Politics," in *The City in American Political Development*, ed. Richardson Dilworth (New York: Routledge, 2009), 1–13, 98–118.

48. Neil Brenner, "Is There a Politics of 'Urban' Development? Reflections on the US Case," in Dilworth, *The City in American Political Development*, 121–40.

49. Richardson Dilworth and Timothy P. R. Weaver, eds., *How Ideas Shape Urban Political Development* (Philadelphia: University of Pennsylvania Press, 2020).

50. Elisabeth S. Clemens, "Lineages of the Rube Goldberg State: Building and Blurring Public Programs, 1900–1940," in *Rethinking Political Institutions: The Art of the State*, ed. Ian Shapiro et al. (New York: New York University Press, 2006), 187–215.

51. Nackenoff and Novkov, *Statebuilding from the Margins*.

52. Jamila Michener, Mallory SoRelle, and Chloe Thurston, "From the Margins to the Center: A Bottom-Up Approach to Welfare State Scholarship," *Perspectives on Politics* (2020): 1–16, doi:10.1017/S153759272000359X.

Conclusion

1. Jessica Trounstine, *Political Monopolies in American Cities: The Rise and Fall of Bosses and Reformers* (Chicago: University of Chicago Press, 2008).

2. John Kingdon, *Agendas, Alternatives, and Public Policies,* 2nd ed. (New York: Longman, 2003). 1.

3. Katherine Cramer Walsh, "Applying Norton's Challenge to the Study of Political Behavior: Focus on Process, the Particular, and the Ordinary," *Perspectives on Politics* 4, no. 2 (June 2006): 353–59.

4. Jessica Trounstine, *Segregation by Design: Local Politics and Inequality in American Cities* (New York: Cambridge University Press, 2018).

5. Patricia Strach, *Hiding in Plain Sight: Cause Marketing, Corporate Influence, and Breast Cancer Policymaking* (New York: Oxford University Press, 2016); Dara Z. Strolovitch, "Of Mancessions and Hecoveries: Race, Gender, and the Political Construction of Economic Crises and Recoveries," *Perspectives on Politics* 11, no. 1 (2013): 167–76; Elizabeth Sharrow, "'Female Athlete' Politic: Title IX and the Naturalization of Sex Difference in Public Policy," *Politics, Groups, and Identities* 5, no. 1 (2017): 46–66.

6. Peter Bachrach and Morton S. Baratz, "Two Faces of Power," *American Political Science Review* 56, no. 4 (December 1962): 947–52.

7. Jessica Trounstine, "The Geography of Inequality: How Land Use Regulation Produces Segregation," *American Political Science Review* 114, no. 2 (2020): 443–45; Domingo Morel, *Takeover: Race, Education, and American Democracy* (New York: Oxford University Press, 2018).

8. Kathleen S. Sullivan, *Constitutional Context: Women and Rights Discourse in Nineteenth-Century America* (Baltimore: Johns Hopkins University Press, 2007); Patricia Strach, *All in the Family: The Private Roots of American Public Policy* (Stanford: Stanford University Press, 2007).

9. "Local Measure A—Garbage Collection and Disposal," June 5, 2012, Consolidated Presidential Primary Election, City and County of San Francisco, Department of Elections, https://www.sfelections.org/results/20120605/ (accessed October 3, 2021).

10. Cole Rosengren, "What's Next for Recology in San Francisco After $36M DOJ Penalty, Bribery Charge," *WasteDive,* September 29, 2021, https://www.wastedive.com/news/recology-doj-nuru-giusti-porter-san-francisco/607153/ (accessed September 29, 2021); US v. Recology San Francisco, Sunset Scavenger Company, Golden Gate Disposal & Recycling Company, US District Court, Northern District of California, San Francisco Division (filed September 9, 2021), https://www.justice.gov/usao-ndca/press-release/file/1431571/download.

11. Rosengren, "What's Next for Recology."

INDEX

Page numbers in *italics* indicate illustrations, tables, and charts.

CPSIA information can be obtained
at www.ICGtesting.com
Printed in the USA
LVHW101031060723
751598LV00014B/171/J